W9-APR-807

THE PEOPLE IN THE PLAYGROUND

The People in the Playground

IONA OPIE

Oxford New York
OXFORD UNIVERSITY PRESS
1993

Oxford University Press, Walton Street, Oxford OX2 6DP
Oxford New York Toronto
Delhi Bombay Calcutta Madras Karachi
Kuala Lumpur Singapore Hong Kong Tokyo
Nairobi Dar es Salaam Cape Town
Melbourne Auckland Madrid
and associated companies in
Berlin Ibadan

Oxford is a trade mark of Oxford University Press

British Library Cataloguing in Publication Data
Data available

Library of Congress Cataloging in Publication Data
Opie, Iona Archibald.
The people in the playground/Iona Opie.
p. cm.
1. Children—Folklore. 2. Children—Great Britain—Folklore.
3. Games—Great Britain. 4. Children—Great Britain—Social life and customs. I. Title.
GR475.068 1993 398'.088054—dc20 92—12172
ISBN 0-19-811265-3

1 3 5 7 9 10 8 6 4 2

Typeset by Best-set Typesetter Ltd., Hong Kong
Printed in Great Britain by
The Bath Press
Bath, Avon

For Sybil Oldfield of Sussex University

Preface

I tried to explain to the people in the playground why I turned up every week and wrote down whatever was happening. (They always call themselves 'people', never 'children'.) 'I think it will be interesting in 100 years' time,' I said. 'I wish somebody had done the same 100 years ago, then we would know what it was like.' 'There was a man called Gilbert White,' I said, 'who lived not far from here, about 200 years ago. He was a naturalist and he made notes of when the children began playing their various games each year, along with his other natural history notes: "1782 28 December, Boys play at Marbles on the Plestor; 1785 23 Jan. Boys play on the Plestor at marbles, & peg-top. Thrushes sing in the coppices; 1788 Feb. 15 Taw and hopscotch come in fashion among the boys."'

The Lore and Language of Schoolchildren had been published in 1959. At the back of my mind was the need to remain in close touch with schoolchildren while we were writing the next two books of schoolchild lore, *Children's Games in Street and Playground*, 1969, and *The Singing Game*, 1985.

I started going down to the playground in the spring of 1960, and in January 1970 began the regular weekly visits that continued until November 1983. My notes gradually became less note-like, and more like a narrative account of what one person could see and overhear and be told directly during the fifteen eventful minutes of morning playtime. I learned to submerge myself as far as possible in the milling throng. I wrote down, in home-made shorthand, the children's descriptions, comments, and opinions exactly as they expressed them, and their stories and jokes exactly as they told them. Then I hurried home to set it all down before my own impression of the experience had faded.

The school is typical of any of the older junior schools in this country. The lore, too, is typical. The games, rhymes, and stories might have been found anywhere in Britain. The only differences lie in the local names and rules for games, and in the deeper, more serious legislative lore, such as the truce term used for opting out of a game, which here in the mid-south is 'Scribs', with crossed fingers as the accompanying sign.

The steep-gabled buildings are standard Board of Education designs. The Infant School, which now accommodates the 7-year-olds, was built in 1872, soon after the 1870 Education Act; and the main school was built on the opposite side of the road in 1914. The classrooms have the lofty ceilings and high windows of their time; and a charming ventilation shaft like a dovecot adorns the roof of the main building. At first the Infant School housed only infants, and the other children walked four miles to the neighbouring market town. When the main building was completed the school could take all ages, though temporary buildings had to be added from time to time until, in 1958, the seniors could go to the new Secondary Modern School in town. A new Infant School opened on the edge of the village in 1973, and the Junior School filled all the old buildings, the 7-year-olds being escorted over to the main playground for playtime. Soon a new junior school is to be built, and the main school will be demolished so that small houses can be built on the site.

I have chosen to present a continuous sequence of entries from January 1978 to the end of the 1980 summer term, believing that this will best convey the reality of the playground; it will also enable individual characters to be followed week by week, and the games seasons to be seen as they come and go. During these eight school terms the average number of children in the school was 250. They were divided according to age into 'first-years' (7 plus), 'second-years' (8 plus), 'third-years' (9 plus), and 'fourth-years' (10 plus).

In our books my husband and I tried to demonstrate the quantity, diversity, and astonishing longevity of children's lore. Now I have attempted to give a picture of the playground itself; to show the lore being transmitted from child to child, the nonce-games being invented, and the behaviour of performers and their audiences. I wanted to recognize that boredom is the bane of humankind, and to exemplify the means children have of keeping it at bay. I wanted, above all, to call up the sensation of being surrounded by the kaleidoscopic vitality of the eager, laughing, shouting, devil-may-care people in the playground.

I.O.

Introduction

1. *The Intruder*

When I appeared in the playground the children wanted to know who I was. 'What's your name?', 'Are you a teacher or a mother?', 'What are you writing down? Can you read it? Can you read it *now*, what you've written?' The girls wanted to know if I had any children, and how many, and, as they were grown up, where did they live, and did *they* have any children.

As soon as they understood what I wanted, they took me on as another chore, something else to attend to during the morning, a more enjoyable duty than emptying the waste-paper baskets. They would come right across the playground to tell me a joke or describe a game. They accepted me as a private individual with the sensible and interesting hobby of collecting games, rhymes, and jokes.

I wanted to experience the life of the playground at its own pace so, when I was not writing down the games and jokes the children brought to me, I wandered around as inconspicuously as I could, looking and listening. My role was passive. I did not dart about asking specific questions and trying to jog the children's memories, as I used to when touring the country with a tape-recorder. If I had done so, I would undoubtedly have collected more examples of traditional lore, for much of it lies dormant in the children's memories awaiting the right stimulus. Nevertheless I was not able to become invisible, and I must have upset the natural balance of the playground by letting it be known that I enjoyed riddling and story-telling sessions; what raconteur could resist the flattery of an adult writing down his story while he told it?

Occasionally a girl would take it upon herself to act as playground hostess, staying by my side all playtime, pointing out events of interest and craning her neck to see what I had written; or an intellectual child, prematurely adult and already surveying the scene as a spectator, would make a few dispassionate comments. Various 'characters' attached themselves. The deprived boy from the Children's Home, who hit the other children without provocation, always wanted me to give him something; one week it was, 'Can I have a bit of your paper?', another week, 'Will you give me a penny?' A neurotically law-abiding boy, who constantly watched to see if the other children were obeying the rules, asked me every time I met him if the headmaster had given me permission to come into the playground. Some children wanted to impress. They made confident (and erroneous) statements about games seasons;[1] or put on an audacious act for me to admire ('I've got a marble-nicking mood on today. It's comparatively easy though. Look, you watch.'). There were also the few unhappy ones, who sometimes came and held my hand as we walked round, because they needed attention. However, the children who came up and talked were the exceptions; most were engrossed in their own play.

At first the playground seemed uncontrolled confusion. Balls whizzed by my head, bodies hurtled across my path, some boys were on the ground pummelling each other, and a dense black mob rushed across, apparently taking no notice of anyone else. Gradually, often with the aid of an interpreter, it became possible to sort out the intermingled games; the chasing game, for instance, which was superimposed upon a diffuse game of Germans and English, both games being intersected by boys competing in running races.

I soon realized that any child with a look of concentration on his face was likely to be part of a game, even though he might be leaning against a wall with his hands in his pockets. I also learned that the children who seemed to be moving aimlessly around on the edge of the playground were usually taking part in amorphous pretending games of their own creation; while others, apparently just 'mucking about', were conducting serious (if minor) experiments on themselves and their environment. A particularly animated group usually indicated the start

[1] Nov 1973. 'The curly-headed girl, who is a television presenter's dream because she is always ready with an opinion, said, "I think skipping's coming back. It's not Jacks [fivestones] any more. It's raining during the day now." "The boys don't seem to mind it being wet for the marbles," I volunteered. "The marbles season is over," she said. "It doesn't seem to be. They're playing all along by the fence." "The marbles season's over," she said firmly, "only they don't know."'

of a new game, with its attendant arguments about rules; though equally it could turn out to be a social incident, such as the explosion of an unexpectedly squashy apple in someone's hand, which showered the bystanders with debris (they make the most of small excitements in the playground).

Children call themselves 'people', rather than 'children'. They say, 'You need six people for this game,' or 'There's some people playing marbles over there.' (Hence my title, *The People in the Playground.*) It is difficult to separate specifically childish words and phrases from adult colloquialisms, but certain usages stand out as characteristically juvenile. Children use 'learnt' for 'taught' ('I learnt the game in Germany, and when I came here I learnt it to Susan and Carol.'). 'I can win him' means 'I can beat him'. 'My best thing' means 'my favourite thing'. They use the preposition 'of' in an unorthodox way: 'I'm bored of this,' they say (taking the construction from 'tired of'), or 'We're having a good time of Kiss Chase.' To 'have' someone is to catch them in a chasing game; to be 'on it' is to have the unpopular role of chaser. If someone breaks the rules in a game they call out, 'That doesn't count!' meaning 'That's not allowed.' 'Guess what [or Hey! do you know what] we've been doing' is a common preliminary to a description of a game. They talk very emphatically, with an extra emphasis on the last word of a sentence ('You don't *have* to do it that way. You can do it the way you *want*.') They use 'just' a great deal, though probably not more than adults do (especially now that 'just' has taken the place of 'only' in, for instance, announcements of bargain prices—'Bedroom Slippers! Just £1.99 a Pair!'). When they tell stories, which they usually do in the present tense, they often say 'he goes' rather than 'he says', a habit which my 17-year-old neighbour has been familiar with all her school life and which, she believes, helps the pace of the narrative as well as being particularly effective with noises and expletives.

The expression most likely to mislead an adult enquirer is 'I made it up.' To a child this is the direct equivalent of 'It came into my head' and has no connection with creativity. For instance, when a child asked me the old riddle 'What did the big telephone say to the little telephone?' 'You're too young to be engaged', and I enquired where she had learned it, she said,

'I made it up. The telephone was ringing and I said to Mummy, "What did the big telephone say to the little telephone?"' The phrase 'I made it up' is so universal in this context that all doubts of juvenile honesty must be suppressed. Probably memory is the same as creation in a child's mind.

2. *The People and Their Behaviour*

The boys come out to play as if ready to fight the world: looking for action, looking for trouble, sometimes turning to spar with each other. A surprising number stand on the top step and cheer before plunging down the steps to the playground. There are always a few who emerge as if shot by a catapult; they run straight down to the end of the playground and, as it were, rebound off the railings, or release their pent-up energy in a wide arc, with arms held out and emitting a long-drawn-out 'Wheeee!' (possibly, though not necessarily, in imitation of an aeroplane). The girls are more sedate. They saunter out companionably in twos and threes, with arms round each others' shoulders, or hand in hand. Boys and girls can be heard deciding what to do (a boy's voice, hoarse and urgent, calls out, 'Let's go and see what that muddy bit's like, come on!'; and a girl says to a friend, 'Going on with Witches and Fairies, right?').

A remarkable number of games can be fitted into the fifteen minutes of playtime if they are minor, transient games rather than grand spectaculars like football and Red Rover. The children play these games rapidly, one after the other, instant decisions being instantly translated into action. A line of leap-froggers quickly turns into a game of Cat's Got the Measles and then again into a hasty bout of TV programmes. If a singing game is broken up by the boys running through it, the girls may not bother to start it up again, but will calmly begin another game instead. After the bell rings they fit in a few clapping games while they wait for the teacher to arrive.

Speed is essential. If they argue about the rules they argue rapidly and agree without much delay, knowing that prolonged argument means that the game may not be played at all.

More often 'the boss of the game' organizes it with force and authority. It is not necessarily the strongest or oldest who become leaders, but those who have self-confidence and the ability to make decisions. Often, by mutual consent, the 'boss' of a game is the one who thought of playing it. The sense of purpose can be felt right across the playground. There is silence while the boss explains the rules to the other players; then she barks her orders and shoves the players into position. It is her febrile energy at the centre of the static group which makes it so noticeable.

In one skipping game observed in June 1970 the boss was a girl called, shall we say, Mandy, and the five other skippers looked to her for instructions. She decided they should play 'Down in the kitchen, Doin' a bit of stitchin', *In* comes a burglar, And knocks you *out*.' Mandy jumped into the rope first. She was certainly a good skipper. The routine went smoothly for several changes, a new skipper coming into the rope at '*In*' and leaving it at '*Out*'. Then someone less skilful came in, stumbled, and had to take an end. Being a less skilful turner, as well, she dropped the rope. There was a restless rearrangement of the group and murmurs of 'Mandy, shall I?' They had a few more turns at this game, then Mandy said they would play a different one. They started skipping through the rope in a continuous line, Building Up Bricks, doing one skip each, then two, then three, and so on. The clumsy 'ender' dropped the rope again. It is not customary to dismiss a player from a game. Mandy had to choose whether to let her continue as an incompetent turner or let her become a skipper again and even more certainly interrupt the flow of the game. She darted forward like a dervish. 'Stop dropping the rope!' she said, picking up the end and thrusting it into the girl's hand. 'We got to three.' (Leaders' faces become fierce and set when they are directing a game—almost evil; yet the expression is only caused by their extreme determination.)

'It isn't always easy to join in a game,' an 11-year-old girl explained to me. 'People are a bit fussy about who plays with them.' I asked a group of younger girls, 'What do you do if you want to join in a game and they won't let you?' 'You go and play the same game on your own somewhere else,' was the prompt response; and she meant 'with your own set of friends'. A solitary child trying to attach herself to a game is an unusual occurrence, and that child is likely to be an oddity of some kind. 'We usually play with our own friends,' they said. 'We usually play in a four, and that is usually enough for a game. If we need any more we go and ask some more people, people we know.' There is a centre ring of friends, and an outer ring who can be co-opted. The four friends come out to play as a group, and are thus insulated against any social problems. They have no need to wheedle or bludgeon their way into other groups or games. 'What if there's a new game that you've never seen before, and you want to learn it?' I asked. 'We watch, and if they say we can't join in we just watch, and soon they tell us to go away.' ('Tell us to go away' gives no notion of the snarling voice and whiplash motion with which this message is often conveyed.[1])

When I asked some boys the same question one of them said, 'You have to ask whoever thought of the game, and if he says, "No," you can't join in.' 'Suppose you very much want to play, can you do any more about it?' 'We just bash them up,' said another boy, bravely but without conviction. The others laughed, and said, 'You just have to go and ask someone else.'

There are many more pairs to be seen in the playground than other-sized groupings, usually of girls the same age; this is especially so at the beginning of playtime before the potato crisps and apples have been consumed. These twosomes may pass the rest of playtime in a domestic and conversational manner (perhaps centring on dolls they have brought to school), or they may play some minor, companionable amusement, or amalgamate with others to play a more structured game. Friendship (which has almost the nature of a contract) begins with two people, and develops into alliances between two couples; further co-operation between foursomes develops according to the needs of a game. Chasing games like Ball He or Stick in the Mud are the most accommodating: 'any number can play', but a typical group consists of twelve to sixteen

[1] Children do not seem to be offended or worried if they are refused entry into a game. The ultimate weapon of telling a teacher would never be used; it is not something recognized as coming within the province of adult arbitration.

children. When a really exciting new game arrives in the playground (as the singing game 'Shirley Temple' did in 1970) the game is paramount; all and sundry join in, girls and boys together.

The convention is that they play with people of the same age. But sometimes an older boy, having become somewhat distanced at the age of 11, will organize a 'horsie ride' for the littlest ones, or an older girl, already maternal, will become the Mother in a game of Mothers and Fathers and lead her 'babies' around on their hunkers. The 'top lot' gave me to understand that they do not as a rule play any games. 'We just stick together, we do. We just walk around and talk.' (Soon after this pronouncement was made, the girls were seen joining—rather shamefacedly—in a game of seeing how high up the wall they could touch.) Some 11-year-olds sound effete and world-weary: 'We find it rather boring in the playground now, and so we stay in and finish what we're doing. We've played all the games, you see. We've skipped and played two-ball and everything for *years*.'

The innate differences between the sexes are easily perceptible in the playground, always allowing for the presence of a few girls who are keen footballers and marbles players, and who are known (and accepted) as 'tomboys', and a few timid little boys who stay under the protection of the older girls. Boys, in general, are more egotistical, enterprising, competitive, aggressive, and daring than the girls. They are comedians, exhibitionists. They do not mind making fools of themselves and provide most of the clowning that is such an important part of playground fun. They concentrate all their attention upon a game. (When boys are in the middle of playing a game it is impossible to interrupt them to ask questions; like an old-fashioned wife, one has to wait for a convenient moment.) They need to be continuously active.[1] They have little regard for their own safety. The fact that they are quite often to be seen crying (whereas I cannot remember ever having seen a girl in tears in the playground) is accounted for by the reckless games they

[1] When a tremendous craze for Jacks swept the playground in June 1973 I asked the boys if they had tried playing. All of them nodded. 'Did you like it?' 'Not enough fun in it,' said one, 'not enough action.'

play, and their propensity for fighting. They are more likely than girls to form gangs or secret societies, and in these their friendship is expressed in terms of loyalty ('If you don't spy on a person when you are told to, you are turned out of the gang.').

The girls' priorities are different. They enjoy talking as a purely social activity, and take far more interest in people than do the boys. They are hospitable and helpful. They crowd round offering crisps, and volunteer to find out about a game going on at the far side of the playground—whereas the boys cannot spare the time. They will stop in the middle of a game to explain it in great detail. They can abandon one game and change to another instantly and without difficulty (which might be called either adaptability or lack of persistence). They conserve their energy, and take care to nourish themselves: many more girls than boys buy crisps at the canteen during break, and wander round eating them when the boys are already playing games. Girls refuse to play some games because they are 'rough—only boys play them'; they are thinking of the damage there might be not only to their persons but also to their clothes ('Our socks might get dirty.').

Some of these characteristics can be seen in a description of girls' games of marbles which took place on 12 May 1973:

> Groups of girls have settled down round the nearest 8″ × 8″ slotted drain covers, eating crisps, with their vanity bags full of marbles on the ground beside them. They are sitting down properly, cross-legged or with legs elegantly out to one side—not crouching and alert like the boys. As I look at them I feel that femininity is a sadly bromidal state, although biologically sound. They scoop the marbles idly from one slot to the next, moving them towards the centre of the drain cover with any one of their fingers, or even their thumb. One of them murmurs to the winner, 'You're good at marbles, aren't you?'; but I feel they scarcely mind whether they win or not. They love the marbles for their colour and texture, and will hold up for admiration a shining ball-bearing, a pearl-white china marble given them by their father, or a jade-coloured glass marble with crystal specks. The boys hover acquisitively. 'I'll give you a game with that ball-bearing in a minute,' says one; but the owner takes no notice.

The girls will go on with their companionable, leisurely games on the drain covers for some time after a marbles season is over.

This rootedness, and the patience they display in games such as ball bouncing, is consistent with the female character—women being the settled sex and continuing longer with a habit.

An obvious difference lies in the need, or otherwise, for an audience. A dozen boys, let loose on the field in the summer, will begin wrestling; a dozen girls will start doing acrobatics. The boys never asked me to watch them wrestling; but the girls always wanted me to watch their contortions: 'Look, Miss, we can do all sorts—headstands, handstands, crabs, cartwheels.' Wrestling is one of the boys' many forms of contest; acrobatics one of many performances given by the girls.

Another ineradicable difference is in their attitude to sentiment. As I was on my way out of the playground one day I saw several girls crouching down by the flower-bed outside the hut classroom. They were decorating a grave. A small wooden cross was fixed upright in the earth. A piece of white paper was laid in front of it representing a marble slab, on which were piled rose petals and leaves. It was the grave of a baby bird. A boy, watching them, commented, 'How thick can you get? The way you're burying that, you're barmy.' The girls turned from him, mildly distressed, and said to me, 'We found six baby birds yesterday. One of the boys picked up one and crushed it against the wall. He killed it.'

Sometimes if a boy finds a dead bird he stuffs it down a girl's back. 'We stuff dead stag-beetles down their backs, too, and squash 'em. They scream their guts out!' This, however, is recognizably a sexual advance.

As might be expected, food and sex are two of the children's principal interests. Sweets are not allowed in the playground, but among the girls potato crisps, in a great variety of shapes and flavours, have almost achieved the status of currency, being handed out to favoured friends, or used to gain popularity. However, sex must rank as the greater driving force. Whenever I noticed any really sparkling excitement in the playground it turned out to be the boys chasing the girls in a game of Kiss Chase (when all the girls have been kissed, the girls chase the boys) or the alarmingly named Snogger ('You chase one of the girls round the back of the buildings, and then you—well, you play Snogger.').

The boys' sexual interest is also expressed by teasing. Running through the skipping games, they try to pull the rope out of the girls' hands, or run in and tread on it. The girls consider this unamusing. They retrieve the rope, wind it up, and move to a quieter place. When the girls play elastic skipping the boys 'come along and ping the elastic behind your legs—it don't half hurt'. It has also been known for girls to seize a football and run away with it, so that they can enjoy the *frisson* of being chased by the boys. As the boys reach the end of their time at the Junior School their attentions become more overt. 'See those girls? They're all—mm—sexy,' they remark, and the girls simper. Little jokes spring up: 'Today's Wednesday—Wednesday's wedding day. If you touch a girl you're going to marry her.'

Fights add to the excitement in the playground, and a fight can be the main, if not the only, event of a morning playtime. Two boys sufficiently angry with each other to start an impromptu boxing match may draw a huge crowd, layers deep in a circle. The cheering eventually attracts the notice of a teacher, who comes and breaks up the contest. If a crowd does not gather the fighting may go on even after the bell has rung:

17 February 1971 After the bell had gone two boys remained. They were unaware of anyone else or of themselves. Their only thought was murder; they were belting each other as hard as they could. A girl who was standing beside me said placidly, 'They *are* in a mood, aren't they!'

But girls fight, too; not so frequently, and in a feminine way with much screaming and pulling of hair—though sometimes in the approved form, with seconds taking messages. I asked a girl to explain a game, once, and she gasped out, 'Sorry, I can't stop, there's a fight on. I don't know what it's about. All I know is that I've got to tell somebody that my sister can flatten her sister.'

One way and the other there is a lot of physical contact in the playground. Boys spar and wrestle and fight. Girls push and shove each other (usually for fun and friendliness). Boys as well as girls walk around with their arms round each other's waists or shoulders; or they mount piggy-back upon each other. Boys who bring forward younger and shyer friends to tell jokes pat them on the head and ruffle their hair. The girls entwine in acrobatic formations. The boys throw themselves at the girls to attract their attention, sometimes organizing themselves into innocent-looking chuff-chuff trains from which the end boy is hurled as a bomb ('We're not just a train, we're a bombing train.').

Apart from its more sensational manifestations, such as snow and frost, the weather seems to have less effect on the pattern of play than does the geography of the playground or the activities of the village street. Although children heat up and cool down more rapidly than adults, they react to weather as individuals and no general statements have been found valid. Certainly it is a myth that in cold weather children run about to keep warm; they are more likely to huddle into their duffle coats, hands in pockets, complaining.

Every feature of the playground is used: the corners and walls of the buildings; the fences (as 'home', or for tying one end of a skipping rope); the ledge outside the largest temporary classroom (for walking along, or as a vantage point, or for a game of King of the Castle); the flat drain covers and slotted drain covers (as sanctuaries or as marbles boards); the small cavities at the foot of 'the marbles fence', where the asphalt meets the grit surface of the lane; the dust-bowl at the edge of the grass, used for flinging toy cars.

Children, like old folk, are natural 'pavement inspectors'. If any technicians start work in the road—refuse collectors, Post Office or Southern Electricity or British Gas workmen—they watch the performance in preference to playing games. Girls hang over the fence 'to watch the ladies go by', and look out for grandparents, mothers, and smaller siblings with whom to pass the time of day.

3. *The Lore*

When *The Lore and Language of Schoolchildren* was published in 1959 it took the intelligentsia by surprise. They were astonished and horrified that children possessed such an extensive underworld culture of their own and were—as Penelope Mortimer put it—'all little savages'.[1] On the BBC's book programme *The Critics* (8 Nov. 1959) the chairman, Nigel Gosling, said helplessly, 'A lot of the rhymes are very stupid indeed. "Horse, pig, dog, goat, You stink, I don't"—well, I don't think anybody could say that was very witty, clever, nor is it poetic.' He had forgotten (or had never known) the necessity for punchy, defensive weaponry in the playground, made powerful by the magic of rhyme. Children's rhymes and games satisfy the requirements of children.

By adult standards many children's amusements are puerile—how could they not be? Children are making first encounters with words, and first experiments in the physical world. They do not invent major new games, all they ever do is slightly alter the established ones or amalgamate several traditional elements. What they do originate are a multitude of 'clever wheezes', hopeful experiments, and minor games which last no more than a day, or at the most a few weeks. ('It's called Needles. You play it when it's just been raining. There are black patches where it's still wet, and white patches where it's dry. You jump from one black patch to another and if you step on a white patch they can pinch you.') I was careful to record such improvisations, knowing that it is an important stage in their development when children contrive entertainments for themselves instead of copying what the others do.

Children also create pretending games, using ingredients from life, books, or television programmes, which can continue evolving over a period of weeks; these they are rather shy about, since they are 'not proper games' and are 'rather silly, really'.

Minor traditional games are the stand-bys of the playground: Trains, Horses, Chiggy-Backs (piggy-backs), Lift Jumps, Follow the Leader, 'A leg and a wing to see the king,' leading-about-blindfold-and-then-guessing-where-you-are, twizzling in couples, and the ever-popular 'Teddy and me went out to tea'. They can be started in a instant and stopped as quickly; they enliven odd moments.

[1] Review, *Evening Standard* (5 Nov. 1959), 14, headlined 'A mother of six looks at a book that will "horrify the nation".'

As well as being picked up in the playground, games and rhymes are learned from older brothers and sisters, from neighbours, and from cousins in other parts of the country; also from parents and dinner ladies. Children live in the present, as a rule; but games learned years ago at previous schools will lie at the back of their memory until some chance circumstance brings them to mind. In the summer of 1971, for instance, Splits (Split the Kipper) arrived in this playground. A girl and boy were playing it on the field, with a steel penknife. She was no good at it; he was an expert. Intermittently, without stopping the game, he told me that he had played it at his old school in Carlisle about two years before. 'And the other day I just remembered it, when I was whittling sticks.'

The kingpins of the playground are the tellers of jokes and stories. There may be only six really good raconteurs in a school of 250 children, and they enjoy much fame and popularity. Their material is very various. 'Jokes' covers almost any form of verbal fun. To children a 'joke' is seldom the kind of miniature comic story told by comedians; most often it is a question with a clever answer that must be guessed, which, if pressed, they would say was a 'riddle'. The 'riddle' may be one of the old, now rare, 'true' poetic rhyming riddles, or a punning riddle ('What key is the hardest to turn?' 'A don-key'), or a conundrum, the most common form of playground 'riddle' ('What is the difference between a warder and a jeweller?' 'One watches cells and the other sells watches.'), or a trick riddle ('Why did the chicken cross the road?') which has a ridiculously practical answer, or a Wellerism ('What did the big chimney say to the little chimney?' 'You're too young to smoke.').[1]

Joke crazes sweep the country from time to time. Most of them last ten years at the most, but a few seem perennial. The craze for 'Knock, knocks', which began in America in the 1930s, is still going strong. The American 'anti-jokes' (or 'flat jokes' or 'let-downs') of the 1960s have not yet lost their appeal. Of these the Elephant jokes came first ('Why do elephants lie on their backs?' 'To trip flying mice.'); then the Colour jokes

[1] Wellerisms, though named after Dickens's Sam Weller in *Pickwick Papers*, have existed since at least 1700. The authentic form of the above would be '"You're too young to smoke," as the big chimney said to the little chimney.'

('What's purple and conquered the world?' 'Alexander the Grape.');[1] then the Lift jokes, which have an element of the Colour jokes in them ('What is green and hairy and goes up and down?' 'A gooseberry in a lift.'). Later there were the 'Doctor, doctor' jokes ('Doctor, doctor, I've only got fifty-nine seconds to live!' 'Just wait a minute, please.'), the cross-breeding jokes ('What do you get when you cross a sheepdog with a jelly?' 'The colly-wobbles.'), and the innumerable other jokes to which the answer is a pun—not always a very clever one.

Children learn jokes from joke books and children's television programmes; they also contribute jokes to joke books, such as the Puffin *Crack-a-Joke Book* (1978), and television programmes. By now, folklorists are familiar with the circulation of folklore into and out of the media, and are reconciled to it.

Some of the long stories which are told with such a sense of achievement follow well-tried patterns. The device of the misleading name ('There were these three boys called Manners, Mind-your-own-business, and Trouble . . .', see entry for 9 May 1978) was used long ago by Odysseus.[2] The 'Englishman, Irishman, and Scotsman' stories of the 1920s and 1930s, designed to poke fun at supposed national failings, are still popular and are sometimes used to ridicule Irishmen. However nowadays, more often than not, the significance of the men being of different nationalities has been lost—they might just as well be called Mr White, Mr Brown, and Mr Green.

In the early 1970s, because of the troubles in Ireland, Irishmen became the butt of a spate of jokes and stories which delighted children as well as adults, and which have now died away. Children relished them not through any anti-Irish feeling but because they always enjoy hearing about ninnies to whom they can feel superior. In fact the foolish Irishmen were part of a long tradition of folktales about simpletons, as exemplified in W. A. Clouston's *Book of Noodles*.

Many of the stories and rhymes boys recite in the playground

[1] See Alta Jablow and Carl Withers, *New York Folklore Quarterly* (Dec. 1965), 250–3.

[2] Odysseus tells the Cyclops that his name is 'Nobody', and when the Cyclops's neighbours ask him why he is calling out in pain he answers, 'It is Nobody's treachery.'

are sexual or scatological. Girls less often indulge in such 'rudeness'. Part of the boys' pleasure comes from the girls' expressions of disapproval; the girls' disapproval is mixed with excitement at the boys' daring. The sexy jokes and stories may not be particularly clever or funny. They are an excuse to use 'rude' words and talk about the forbidden topics which are risible in themselves. Defecation, urination, nudity, the private parts, the sex organs, and the sex act, which, in a civilized society, are kept private, have a fearful fascination for children. The growth of permissiveness has meant that fathers and older brothers now share sexy stories with the younger boys, and children stay up late to watch extremely 'adult' programmes on television.

Children already seem to know that 'human kind cannot bear very much reality'. Childhood is a time more full of fears and anxieties than many adults care to remember, and play is the way of escape. A game is a microcosm, more powerful and important than any individual player; yet when it is finished it is finished, and nothing depends on the outcome.

Try to analyse the sound of children at play: the thin screaming noise can be heard from several streets away. Vitality? Yes. But come closer and step into the playground; a kind of defiant light-heartedness envelops you. The children are clowning. They are making fun of life; and if an enquiring adult becomes too serious about words and rules they say: 'It's only a game, isn't it? It's just for fun. *I* don't know what it means. It doesn't *matter*.'

The playground: north side (above) and west side (below) with 'tall ball wall'. This side of the playground, being quieter, is also where the girls play their ring games.

Thursday 5 January 1978

Term started yesterday. The weather is bleak (this is not, so far, Sara Coleridge's January of snow and glowing fingers).

The children marched across the road from the old Infant School, bringing cheer into the bare, cold playground. As they came I heard the born organizers at work, calling to their friends, 'Come on, we'll play ——,' 'I know what we'll do, we'll ——.' But as it happened there was an entertainment available more attractive than any game. Southern Electricity men were replacing a bulb in one of the street standard lamps, and most of the children went to watch. A man in green overalls was lifted to the light on a hydraulically operated platform, a circus stunt in itself. The market gardener's son remarked, 'Did you notice, they use an ordinary light bulb—at least it looked like an ordinary light bulb.' (He is a serious boy, worth talking to.) A chubby blonde girl said, 'That green man looks like Rolf Harris.' (Rolf Harris is a well-known television personality, I know that much.)

The crowd began to disperse, though seemingly in no hurry to begin playing games. A girl said conversationally, 'I got a New Year card from my friend in Wales. She sent me a Christmas card too.' 'I got a skate board for Christmas,' said the slender worldly-wise boy (curly hair and knowing eyes). 'I skate on the road, I live up that turning by Milland's.' The market gardener's son said, 'My mother's going to buy me a skate board when they make a proper place for people to skateboard on.' 'He's going to have a home-made one,' said the first boy. 'No, I'm not,' said the other bravely, 'I'm going to have a plastic one.'

Two kinds of people inhabit the playground. The kind who, on a cold day, go into hibernation, retreating into their hooded overcoats. And the other, more numerous, kind, who manufacture excitement which, to a rational observer, is unnecessary, even silly; yet this manic activity is probably more profitable on a cold day, if one can subscribe to it. Cohorts of boys were goose-stepping round the playground, shouting, 'We are the

Trouble Makers!' and trying to kick everyone within sight. Flocks of little girls ran by, screaming, 'Help!' Chasing games of a more conventional nature were in progress too. A resolute Jack the Giant Killer, aged 7, chanted 'One, two, three, four, five, six,' before setting out; and a girl pursued her friends with a Kermit the Frog 'Muppet' puppet, pulling up her red socks with her spare hand.

'Do you want to see our horse when it's ready?' said the flat-faced girl, and assembled a boy and girl into an animal shape behind her. Another girl leapt astride the animal, yelled, 'Gee up!', and brandished a make-believe whip. 'I thought that is usually a camel,' I said. 'Well it's like a camel. With a camel he would have to stick his bottom up,' said the girl, slapping the middle player on the behind.

I waved to Mr ——, on playground duty in his duffle coat, his nose pink and face pinched with cold. A few boys were playing marbles for want of a better idea, but they were chiefly concerned by a marble stranded in the lane. 'Can you fetch it?' Paul's brother asked me. 'It's a cat's-eye.' 'What kind is that?' 'Well, it's green and see-through, and it's worth a lot of marbles.'

I was glad to see people exerting themselves, even if not effectively, in this pale season of the year. All over the playground were single examples of games (one skipping rope, one hand-clapping game, one game of Sly Fox) as if the children were trying to recall what games there are to play. Three were huddled in a corner with an old paperback Animal Quiz Book. Everyone seemed relieved when the bell rang.

The seniors hailed me as a diversion while they waited for their teachers to reappear. I felt almost as popular as the Road Safety Clown. It was a question of who could get in first with his joke. The recognized story-tellers have the advantage, and one of them began. 'There is these three ducks, right?' 'Is it rude?' said someone, and another said, 'I hope so,' but the story-teller did not hear them. 'The first duck went, "Quack, quack" [loud and plain]. The second went, "Quack, wack" [more faintly]. And the last went, "Wank, wank" [more faintly still].' They were all hugely delighted. The market gardener's son said, looking at me earnestly with big brown eyes, 'What is the last thing you take off when you go to bed?' 'Er, your vest?' 'No, your feet off the floor.' A neat exchange, giving intellectual rather than Rabelaisian pleasure.

The story-teller, who lives in a constant state of geniality, was prancing round me clutching a plastic bag of biscuit crumbs. 'You ain't having none of these,' he chortled. 'Oh, where did you get them from?' I asked, forgetting that he was not a fellow housewife. 'Up the common,' he said, 'when I was digging for food.' 'You'd better go in,' I retorted unfairly, 'Mr —— arrived long ago.'

Friday 13 *January*

China-blue sky and stinging wind.

The first children came out, ran to see why one of the drain covers had been removed, and stood gazing at the brimming yellow water. The caretaker went up to them and said, 'Will you go over to the other side of the playground, please? This is not for you. It's not very nice 'ere.' They went.

More of them came out and quested, as dogs do when they are let out in the morning. They found that an oval of ice had formed under the drinking fountain; the boys hung on to the fountain and slid on the ice as if they were working a treadmill. The girls stood and watched, eating their potato crisps. The dainty girl with the fringe and blue eyes offered me a crisp. 'Roast *chicken*,' she said succulently, the tone somewhere between that of a persuasive mother and a television commercial. We walked away together, social contact having been made. 'I like your ear-rings,' I said. They were tiny silver butterflies. 'I got them for Christmas,' she said, 'and I had some pearl ones like yours.'

We strolled on until we reached the end of the big wall, where girls were sitting in the sun playing with their Christmas dolls as if it was a day in June. They had established a home on three yards of asphalt. 'She's going to bed now,' said one, as she put a teenage plastic doll into a plastic double bed; and another, her teeth chattering, concentrated on fastening a pale blue brassière on to a pink chest. Round the corner the boys were sitting in the sun, too, swapping car cards ('They're called Prototype, this kind. You can get all different kinds.'). 'They're not allowed to play football any more,' said my friend, *sotto voce* so as not to hurt their feelings. 'They broke a window. You can see the paper stuck over it, look.'

'What's all that whitey stuff on the ground, little chunks of whitey stuff?' I asked the boys. 'Is it glass from the window?' 'It's ice,' they said. But I went to see, all the same. 'It's ice,' said the boys who were over there. 'You spell it I-C-E. The girls are getting it off that puddle. You want to hear a joke? What is a sick kangaroo? A Hoperation.' 'Knock, knock,' said another boy (the tough apple-faced boy from last week). 'Who's there?' I dutifully replied. 'European.' 'European who?' 'European down your trousers.' 'Oh, that's a good one,' I said, knowing he would be pleased at the praise, yet thankful it was said impersonally, as if he scarcely knew what it meant. (You can hardly believe your ears when the kids tell you these jokes. He did not sound as if he was telling a rude one.) 'Knock, knock,' he said again. 'Who's there?' 'Catch up.' 'Catch up who?' 'Catch up with your homework.' 'That's not such a good one,' I said stoutly, following the principle that honesty is good for the soul. 'I think someone must have made that up.' 'I did,' he admitted, a little dashed. 'Well, score one for making it up,' I said penitently. 'I expect you were thinking of catchup being a name, like tomato ketchup?' 'Yes,' he said; but I am not certain he had thought at all.

A few girls were skipping (one game of High Jump, so called) but otherwise I could see no concessions to the biting cold. The children were, for the most part, sitting, or scrimmaging idly, or standing chatting, or experimenting with the ice. The flat-faced girl joined the line by the 9-year-olds' classroom, after the bell went, and told me with some annoyance, 'We've been smashing ice on the ground to make a slide, but it didn't work. Every time you ran across it, nothing happened.'

My fingers, inside woollen gloves, were quite painful. I asked a little boy swinging a yellow towelling marbles bag, 'Wasn't it cold playing marbles? Aren't your fingers cold?' He hardly knew what I meant. '*No*,' he said, staring at me, and went on swinging his marbles bag. The teachers, however, were not inclined to stand about once they had arrived. Their voices had an extra briskness as they called their classes to order, and I tactfully left as soon as I saw them standing on the top step.

Tuesday 17 *January*

Damp and cold. It would have been better if Sara Coleridge had written: 'January brings the dirt, Makes our feet and fingers hurt.'

The playground was steel cold, and empty except for a sparrow hopping between a flattened aluminium milk-bottle cap and a crumpled Crunchie paper. I saw children surging towards a classroom exit, and went to the steps and stood there while they came past me, arranging their games. 'First-years against second-years, OK?' 'I know, let's play Om Pom [hide and seek].' Two girls approached a third and said with almost oriental politeness, 'Melanie, may we play Stuck in the Mud, please?', and Melanie gave them permission. I was startled into asking her, 'Are you in charge of them, or what?' She looked slightly puzzled herself. 'Well, I think it's that normally I think of Stuck in the Mud.' The incident emphasized the natural tendency children have to be law-abiding. For instance, the law forbidding the whipping of tops in public places has long since fallen out of use; but children will say in awed voices that tops are illegal, and so of course they cannot play with them any more.

The flat-faced girl, interested in natural philosophy as always, had scratched the heads off a number of acne spots on her face and was mopping the blood with a handkerchief. 'You've got a mess there,' I observed. She looked at the handkerchief in a detached manner and said remotely, 'Yes. I got the measles.' Such remarks will infuriate her husband one day. They are beyond the reach of human repartee.

Two 8-year-olds were trying to recruit players for a game of skipping. As they turned their too-short rope without enthusiasm they asked each girl who came near, 'D'you want to join it?' The girl with the dark fringe and deep-set blue-grey eyes was sure she did not want to skip. She brushed them aside as if they were collecting for charity and came straight to me, saying, 'I've got a joke. There's this Englishman, Scotsman, and Irishman, you see, and they all went up the Eiffel Tower, and the Irishman was short of money and he said, "I bet you £20 you can't drop your watch and run down and catch it." So the others said, "OK." So the Englishman tried, and he failed. And the Scotsman tried, and he failed too. And the Irishman dropped his watch, and he ran down, and he had a drink, and he went for a drive in his car, and he had a bath, and he went back and caught his watch, and the others said, "How did you do that?" and he said, "I put it half an hour slow."'

There was a vague mêlée as the audience moved away (one is never sure how the grouping changes in the playground: the children coalesce and disperse like clouds in the sky). I found myself facing a primitive figure in a worn duffle coat. 'I wondered what you are playing,' I said in the dream-like voice I often use, as if I was part of their game already. 'We're playing Anti-tanks and Anti-aeroplanes,' he said. 'We're killing elephants and the warring mammoths. We're trying to kill them, anyway. That's how they got extinct.' His small face looked desperate inside its sackcloth hood. He seemed to have stepped straight out of Thornton Wilder's *The Skin of Our Teeth*, and still to be worried by the besieging mammoths.

In the distance a boy was firing two bright yellow bananas at a distant target. The girl who possesses many pairs of ear-rings drifted up and said dolefully, 'I had a bad ear last night, all bloody. All because of the hole for my ear-rings.' She gave me a pathetically feminine look and drifted away again.

'Miss, Dave wants to tell you a story.' A rather responsible-looking girl brought me the boy who tells rude stories, with the announcement, 'This is Dave,' and left him to tell his story. 'This man went into a garage to get two drops of petrol. 'E says, "'Ere, 'ave you got two drops of petrol for me?" and 'e gives 'im two drops of petrol. When the boss comes back the man says, "What about this old geezer? 'E comes in 'ere and wants two drops of petrol." And the boss says, "Oh yes, that's old Bert. I'll go and see what 'e's doing." Well, 'e goes to the toilets and this old geezer is in the third cubicle and 'e's going "Errrr" [making the sound and imitating the action of a man starting a motor]. So the boss goes back and 'e says, "It's all right, old Bert put the drops of petrol on 'is machine and 'e's going like this: 'Errrr' "—and 'e shows the other man.' By this time, having repeated the engine noises and the engine-starting motions somewhere in the region of his trousers the boy and his audience were heaving with laughter, and I suppose even I did not really need his shouted finale, full of heathen joy—'It's a wanking machine!'

Paul's brother was waiting to meet me round by the senior classrooms (the bell had gone). 'I was wanting to tell you this story,' he said seriously. 'There was this man, OK? and he was in love with this lady, and 'e 'ad it off with this lady, and the lady's husband come home—' Some of the other boys were beginning to laugh. Paul's brother rounded on them. 'This is really true, OK?' he said fiercely. 'Yes, it was in the papers,' one of the boys said, supporting him. 'Well,' said Paul's brother, ''E 'ad it off with 'er, and the husband come home—' 'I heard that one, it was ages ago,' said another boy. Paul's brother did not take any notice. 'And the husband cut the man's willy off, and the police come and they had to go looking for it.' 'Yes, that's right,' said a boy. 'It was in the *Sun*.' It was worth waiting to hear that story. It is possibly related to the so-called 'urban legend' currently going the rounds, in which a little boy's willy is cut off by immigrants in a public lavatory.

Thursday 26 *January*

Pale blue sky, windy. The playground covered in rivers of melting snow.

The boys came out with their heads down and hands in pockets, stamping in the slush and gauging its possibilities. 'Here, this is the best ice,' they cried. The headmaster appeared on the school steps. 'Go in the clear part of the playground, please,' he called out, 'not here where it's mucky.'

We all had to move round to the south side, where the senior boys were already shovelling the wet snow into the drains. A 7-year-old boy was standing as if waiting for me, while a little girl tied up his shoe-lace. 'I'm not doing anything this morning,' he announced, and indeed it seemed to be true. 'Can't he do up his lace himself? Is he your brother?' I asked the girl after the boy had walked away. 'No, he can't. I haven't got any brothers, thank goodness,' she replied. The incident, and the girl's old-fashioned, new-fashioned clothes (a Victorian fur-trimmed and hooded cloak), reminded me of a family story. My father, sent home from India at the age of 7, had to have his boots unbuttoned by a girl cousin his own age, and never forgot the shame of it.

The small bouncy girl started jumping up and down and flapping the ends of her sleeves. 'Look, I'm a Billy-wobble!' She ran and fetched a friend. 'This is Helen, she thought of it.' Helen peered at me from under the hood of her quilted anorak, and ran away again. 'She says they live in Greenland. They're really penguins.'

A boy planted himself in front of me, a stranger, and without any preliminaries began, 'There is this Englishman, Scotsman, and Irishman, and they met on a beach at a cave called "Echo's Hollow". And the Englishman went in the cave and he saw there was a pound note on a table and he went to put it in his pocket, but he heard a voice saying:

> I'm the ghost of Aunty Mabel
> That pound note shall stay on the table.

Then the Scotsman went in and he saw the note on the table and went over to put it in his pocket and he heard the voice saying:

> I'm the ghost of Aunty Mabel
> That pound note shall stay on the table.

Then the Irishman went in and he actually put the pound note in his pocket, and he said:

> I'm the ghost of Davy Crockett,
> This pound note shall stay in my pocket.'

It seemed very funny at the time. It was a social success.

I said, 'I *must* go and see what games people are playing,' and went to look at the only game to be seen, which was marbles. The boy followed. Devotion to marbles has remained steady for several years, perhaps because there has been no feverish craze during that time to burn up the enthusiasm. It has been quietly available as an alternative amusement. This morning boys and girls were spaced out along the fence, playing with china marbles (now the best buy) and fishing the marbles out of inches of water with the toes of their shoes.

'You haven't heard this one,' said a determined voice at my side. 'It's about this man and he was going to jump off—was it the Tower of Pisa? no it was the Post Office Tower. This lady she went to get the priest of the Catholic Church and she says, "There's this man going to jump off the Post Office Tower," and the priest says, "I'll come and see him," and the priest gets up to the man and says, "Don't jump. Think of all your family and your wife," and the man says, "My wife has left me and I have no children," and the priest says, "Think about the Catholic Church," and he says, "I'm not a Catholic," and the priest says, "Jump then!"' He told it, as he had told the other story, with an expressionless face and looking straight ahead. I could see he was just able to hang on to the thread of events by gazing into the distance and not allowing himself to be distracted; but he was able to stop when I said, 'Just a minute,' and start again where he had left off. He used the present tense, as do all the children, and occasionally got a phrase round the wrong way (for instance I imagine the original wording was 'Think of your wife and family'). 'Where did you hear that?' I asked. 'On the telly?' It had sounded like a Dave Allen story.[1] 'Yes,' he said. 'Not Dave Allen though. I forget.' While he had been reciting the story with such concentration I thought, 'He is learning about life on trust from television, and is repeating his lesson.'

When the bell rang I went to join the seniors. The snow-clearing team had reached a riotously playful stage. They were tilting at the drains with their shovels and brooms. The bold bad story-teller caught up with me, brandishing a length of joined elastic bands and saying, 'This is a threatening machine.'

[1] The sophisticated Irish television raconteur.

'I really need some more games,' I said. 'What about Shagging? That's a good one,' he said. He has glowing honey-amber eyes, and an air of radiant health. He is the kind of healthy animal one would forgive for anything, as the Bishop in Norman Douglas's *South Wind* forgave his healthy savages their lapses into cannibalism. 'What about Penis?' said the boy. 'That's a game. Yes it is. The girls chase the boys and when they catch them—Oh, I know a story. It's about an old woman and she had a daughter called Fuckerard.' He had a moment's doubt and asked his mate to 'ask her if she writes down muck and all that'. I said, 'I write down whatever's happening.' 'Oh good. Well, there was this daughter, and the old woman went out and the bread man or the milkman or somebody comes and he says—You don't show this to the teachers, do you?' 'No, of course not.' 'Well, the milkman comes and 'e says, "I'll give you two pints if I can come inside." So 'e comes inside. Then 'e says, "I'll give you two pints if I can come upstairs." So they get upstairs. Then 'e says, "I'll give you two pints if you get into bed and get undressed."' ('You got that the wrong way round,' hooted another boy.) Mr —— went by, smiling. 'I'd better go,' I said. 'Oh no, there's only a bit more. The old woman comes back and she comes upstairs and she comes into the bedroom and she shouts out, "Fuckerard!" and the milkman says, "I'm trying."' 'No, that's not right,' said the other boy, 'He says, "I can't fuck 'er any 'arder or I'll go through the bed."' The story-teller let his breath out like a stuck pig. 'Whew!' he said, and I knew how he felt. They ran off after Mr ——, one saying to the other with deep appreciation, 'Oh I *like* that one.' It is true that sex is the basic life-giving force.

Wednesday 1 *February*

Chill grey moisture-laden air.

The senior boys jumped down the steps and ran to the marbles fence. One of them put two marbles into a hole and shouted, 'We lay! We lay!' 'What does that mean?' I asked. 'It means me and Colin are going to play marbles. It means nobody else can't play in that hole.' Marbles is suddenly serious, highly competitive, and commercial, sparking claims and counter-claims. 'Touch!' cried one. 'Stops!' cried another, putting his foot on a marble. I felt as bewildered as an American at a cricket match. A girl standing next to me offered to explain. 'Touch—well, if you don't want to do anything you say, "Touch!" and touch it and that takes up one turn.' She had a tin of marbles and I thought I might be keeping her from her game. 'Are you playing, or are all the holes taken?' 'Yes. There are some little ones but they're not much good.' A friend came and opened the lid of the marbles tin. She routed about with a finger. 'Where is it? Where is the big bull-bearian?'[1] 'It's inside,' said the first girl. 'Shall I go and fetch it?' She went and fetched a large scratched ball-bearing from her classroom and gave it to the other. 'She's going to play it for me,' she said, and resumed her role of mentor.

'This game, you see, they're playing eight coloured fourers for a blue see-through.' The see-through was a rich Bristol blue, rarer than the ordinary glass swirl fourers. There was treasure at stake. The spectators were working up some excitement, too, exclaiming, 'Gosh!' and saying, 'Boo!' unexpectedly to put the aimer off his stroke. 'No drags!' said one player. 'That means you mustn't push it along with your finger,' said the girl. Someone called, 'Aims away!' '"Aims away"? Well, it's like having a point, almost, having an "Aims away". Say I had the last marble in the hole and I had an "Aims away", I would have won the game.' 'What's the other thing they're shouting, "On the line"?' 'That's if your marble stops on the edge of the hole. You have to chuck it out—I mean you have to send it right away.' The game was over, the blue see-through had been won. With one movement the winner scooped the marbles out of the hole and threw them a distance away ready to begin a new game, calling, 'No stops!' The next game began from where they came to rest.

A hefty lad with bright eyes and curly hair had been standing near me twirling a heavy metal pistol round his finger, cowboy fashion. He seemed pleased to be asked what kind of weapon it was. 'Quite an old one, yeh. A Luger. Blows everyone's heads off,' he said complacently. Then we were surrounded by small

[1] Local pronunciation of 'ball-bearing'.

Charles Addams people, with round vacant eyes and grey faces, stranger by far than cowboys.[1] They had their thumbs in the sides of their mouths and their teeth over their lips and they said, 'We are the Martians.' 'Have you heard of this stuff called rain?' one said, holding a hand up into the drizzle. 'Look, here are our radio sets,' and they flipped open notepads with radios drawn on them. 'I'll show you, I'll show you how a Martian talks,' said one, opening another Martian's palm (it had '2' written on it). 'What's your name?' he asked it. 'Stephen,' said the Martian in a funny voice. 'Where do you live?' 'Down the bog.' 'How do you make love?' 'Kissin' and cuddlin'.' ('Shagging,' corrected another boy.) 'Ah,' said his interrogator, 'we make love by drawing little circles round and round people's hands,' and he tickled the other's palm as if playing a Martian baby game. 'Hey,' said the loved one, thoroughly roused, 'I'm having a drink of lemonade—pissss. And a piece of chocolate cake—whee!' He danced about, plucking at his trousers.

The rain began to come down in earnest. The children took no notice. Even after the bell rang they stood in the downpour pressing me to listen to stories about Englishmen, Irishmen, and Scotsmen. 'I've heard those,' I objected. A fur-collared cad said, 'Well, what about this? We went into Portsmouth on Monday, to see Pompey play. We skived. And we went to a strip club. The lady in the strip club said, "D'you want a fag?" I said "Yeh, but I want something else, I want your telephone number on a bit of paper."' He is going to give some girl a thrill one day.

Wednesday 8 *February*

Grey sky, face-biting wind.

The mothers and fathers of Hamelin Town could not have felt more desperate than I did when I saw Mr —— leading the senior children away a few minutes before playtime. However, I went to see what a depleted playground could offer.

[1] Charles Addams, *New Yorker* cartoonist, creator of the sinister Addams family.

A stocky, trim girl came up with elastic step. I searched for adjectives to describe her, and found myself thinking in terms of nationality—Dutch, Scandinavian, German—to pinpoint her characteristics: something to convey the pale face, plaits, air of toughness. Her white lacy tights gave her a gymnastic rather than a dainty look. She struck an attitude and said, 'I've been away all this week with a bad cold. I should have put a coat on but I don't want to—want to get the fresh air.' A young teacher walked by on her way to have coffee in the staff room. 'Karen, you put your coat on,' she said firmly; but wisely did not wait to see if Karen obeyed.

'Karen, come and hold an end,' called a girl who ran out of the school door carrying a long skipping rope. Karen and I went with her to the far side of the school. The other end of the rope was given to a strangely elderly girl, fleshy in a Spanish style, who proceeded to turn it with small wrist movements, like a dowager unexpectedly asked to conduct some chamber music. The method was surprisingly effective. The owner of the rope, Vicky, was of course not one of the enders. She organized and supervised the game while taking part in it. She was keen that it should work as smoothly as possible and, herself of elfin build, lambasted some girls who were slow on their feet and had not, it appeared, even been invited to join in. As she ran round the ender for her next turn she scolded these players: 'Sharon, you must *not* go in,' 'Tracey, you keep out, you weren't asked.' She had time to complain to me over her shoulder, 'They keep slowing up the rope when they try to jump.' She had my sympathy. The game was Building Up Bricks. In this game the players have to skip one, and run out of the rope, until all have had a turn, then skip two, then three, and so on; speed and nimbleness are of the essence. The impresario's task was made no easier by the little boys who were playing football with a glass marble in and around the skipping game.

I found myself talking to three boys who had lined up in a row with bright morning faces. I said, 'Not many people in the playground.' 'No, everyone's got chicken-pox.' 'My sister's got flu,' said the largest, and whispered to either side. 'It's rude to whisper,' I said in the gleeful now-I've-been-given-an-opening voice a child would use. All three proffered a crisp at the same time. I helped myself to a KP Outer Spacers chutney-flavoured four-pointed-star crisp from each bag (and very energizing they were). 'That's what I was whispering about,' said the largest boy, cock-a-hoop at having caught me out. We chewed our chutney crisps. 'Just bought them from the canteen,' said the big lad. 'Cost 5p.'

By the wall a short string of girls were calling 'Join in, please,' collecting people to play Sly Fox, and before I could reach them the game had started. I leant against the wall near the one who was 'on'. The girl at the wall was scarcely giving the others time to move half a step. 'See you, Debbie. Your legs. Go back to the start. See you, Paula. When you were like that—your legs moved. See you, Maria—' She had to give them evidence that she saw them moving. All the same, so many were playing they reached her from the side she was not surveying, touched her, and began the long chain for others to touch. Soon all were home, and she started to tell the story. 'One day there were—[she counted them]—ten princesses.' 'Oh no, not that one,' groaned one of the players. 'It doesn't matter which one you have. One day there were ten princesses and their names were Princess Tracey, Princess Debbie, Princess Maria . . . and they lived in a big castle and they had a big garden. One day they went in a wood and they came across a sly—*fox*!' The girls ran back to the starting line, with the story-teller after them. As she chased them she shouted, 'And they met a prince, and he loved all of them.'

While I had been leaning against the wall child-life had been going on all around, almost rubbing against me. Two little girls scuffled in the corner, apparently taking turns to smother each other with their coats. A girl leant casually beside me, one foot against the wall, playing cat's cradle as if it were mid-summer ('Well, it's not cat's cradle actually. I'm making a parachute.'). A boy repeatedly kicked a red plastic folder or pocket book against the wall, scoring goals. When the bell went I joined him by the gate. 'It is a case for keeping coloured slides in, really,' he said, and showed me the colour trans-parencies inside. 'Oh,' I said, 'doesn't it spoil them?' and remembered too late that it is not for me to be censorious. I said, 'I suppose you are still not allowed footballs.' 'No,' said another boy. 'People use anything, gloves and that.'

The seniors were coming back from over the road. 'We were

watching a television programme about jobs,' they said. 'About building sites, and carpenters, and things like that, and about the noise they make. I expect we'll be given some playtime, we usually are.' I followed them into their cloakroom, for warmth. Two of the boys jumped on a shoe locker and looked down at me benevolently. (I think they were the Plymouth Brethren.) 'Try putting your fingers in your mouth and pulling it sideways, and then say, "I was born in a pilot ship." It comes out "I was born in a pile of shit."' They were interested in the mechanics, almost entirely. 'You can say, "I'm not a plucker, and I'm not the son of a plucker", too.'

They were not anxious to go and play. The playground does not seem exciting when only a third of the children are out in it. A girl wanted to tell me about her 4-year-old sister. 'She said she wanted to give my Dad two sausages for his birthday, and she did. My Mum wrapped them up in paper and everything.' She beamed at me and we shared the naïvety of a 4-year-old's predilection for useful presents.

I skirted the small crowd of marbles players and decided playtime was over as far as I was concerned. Then I remembered I had wondered whether these children know the 'Jingle bells, Batman smells' rhyme we are being sent from other parts of the country. They didn't, but two of the girls ran after me to say, 'We know a rhyme, but it's rather rude.' They recited:

Marty Farty
Had a party
All the farts were there.
Tooty Fruity
Did a beauty
So they all went out for air.

Such nice girls. No amount of low company alters their standards of 'rudeness'.

A boy came up just as I was leaving, to say, 'I know that Batman rhyme. It goes:

Jingle bells, Batman smells,
Robin's flown away,
The Batmobile has lost a wheel
And landed in the hay.'

Monday 20 *February*

Snow still a couple of inches deep. I went down to the playground today, the first day after the half-term holiday week, to see what the children would do with the snow. The playground was an unblemished snowfield, through which the caretaker had swept meandering paths. The temperature is above freezing, and the paths were wet. The school secretary said the children would not be allowed out: 'There's a lot of snow in the playground which might get trodden into the school.'

Tuesday 28 *February*

Strangely warm after the snow of last week. Birds cheeping, buds burgeoning: a feeling of energy returning. All the same it was not a beautiful spring day, it was spitting with rain.

I walked straight into the playground, eager to get at the children after the frustrations of last week, and was greeted by a small joker with '*You're* not here again, are you?' (I thought this indiarubber jollity was something one acquired as a protection later on in life.) 'Gotta go and play football,' he said. 'It's only a *kind* of football, with a marble, on a drain cover.'

'We've just remembered a game! Shall we show you?' said the sociable lot. 'It's called Sorry, Sir.' They folded their arms and bumped each other indiscriminately, saying, 'Sorry, Sir,' each time they did so. 'You don't have to bump each other in any special order?' 'No, you just bump anybody—anybody that's playing, that is.' 'I used to play it at my last school,' said the tall pretty girl, 'and the others knew it already.' 'We *didn't*,' they said. 'Well, I just thought of it,' she said. She and her sister have been at this school for two years. Although it might well be a Scout game, she said she had always played it in the playground.

The flat-faced girl came to pass the time of day. She is so tall and burly, not to say mature, I had not realized she is only 8 years old ('Nearly 9—my birthday's in March'). She stands confidently, sure of her place in the world. I asked her, 'What have people been doing?' Her pale blue-grey eyes looked straight into mine. 'They've been playing football with little tennis balls and gloves and things. And some people have been playing Lines—chasing on the netball lines, you know.' A rather strange-looking boy with peeled-back eyes said, 'Some girls have been playing hopscotch, although we're not allowed chalk,' and then became confused when I tried to find out how they played—he thought it might have been on the lines of the netball court, after all.

The girl said urgently, 'I've got a joke for you. How do you know an Irish pirate? He's got patches over both his eyes. Do you want another Irish one? There was this Irishman and he threw his hat on to the ground.' She laughed up at me. 'He missed,' she said. 'I don't understand,' said the boy. 'I got another one,' said the girl. 'Doctor, doctor, I feel like a snooker ball. Well, get to the end of the queue then. And there's one about an orange. Why did the orange go to the doctor? Because he wasn't peeling very well.'

We had settled into a temporary community, the jokes as usual having attracted an audience. A boy wearing a Planet-of-the-Apes mask clattered his plastic teeth. Masquerading children seldom say anything. 'What about Star Wars? Can you get Star Wars masks?' I asked. A boy said, 'You can buy Star Wars lollies and you get a mask free. My sister's got one.'[1]

When Mrs —— arrived I asked her if the children are really not allowed to use chalk in the playground. 'No, they're not,' she said. 'The staff did some drawings on the wall, too, and Mr —— said we must stop. But they wash off, don't they?' I had to suppress an alluring vision of the staff expressing their personalities in graffiti. Mrs —— saw the light in my eye and said hastily, 'It was for outdoor games.' Then I met the school secretary and asked her about the children being allowed chalk. 'No, they're not,' she said firmly. 'It's funny, because there was this marvellous hopscotch craze in 1974, and they were doing it last summer as well. Perhaps they're allowed it just for

[1] *Star Wars*, the huge box-office success from Twentieth Century Fox Film Corporation, 1977.

hopscotch. Perhaps they had some special chalk that washed off. I don't know really. I'll ask the headmaster. He's not here today.' I realized she had been left as watchdog of the school, and hearing the telephone ring I decamped.[1]

Thursday 9 March

A nasty grudging day with a lid of grey cloud and a sneaking wind. I thought the children were going to be withheld from me, but my watch was fast.

The slim young teacher of one of the second-year classes came out of the main door and walked briskly across to her classroom, efficient in her pencil-slim check skirt, green stockings and medium-heel county shoes. I felt excluded. Suddenly a boy pelted across the playground, looking over his shoulder to see if his marbles opponent was following. They bent to their game straight away. The first boy threw eight marbles up to the hole, one at a time, getting enough in the hole to win extra turns to edge the remainder in. But why start headlong with eight? 'He lays all of them he wants to play for mine and if he gets all of them in he has first shot,' said the other boy, and then addressed a larger boy who had draped an arm over the railing, 'Got any marbles?' 'No,' said the larger boy. 'You should know that. You won 'em off me.' I was worried about this shooting of eight marbles; it did not seem familiar. I asked the bankrupt spectator, and he said, 'Yes, usually you send one up and the nearest to the hole has first shot. But if you both get them in the hole you take them out and shoot again.'

Two merry little girls with perfect teeth stood in front of me

[1] When I met the headmaster out shopping I asked him about chalk being allowed in the playground. He seemed surprised and said, 'I don't think any decision's been made. I haven't said anything.' Then he said, 'They were playing hopscotch last summer, weren't they?' I ventured, 'The children are quite sure chalk is not allowed. Children are so law-abiding.' 'Yes,' he said, pleased, 'especially in this school.' 'A teacher told me the staff drew some diagrams on the walls and you asked them not to.' 'Yes, and even then she hasn't got that quite right. What happened was, I said "Don't draw on the walls with chalk. If you want diagrams I'll have them painted".' We agreed that no one ever got anything quite right. The worry seems to be that chalk marks in the playground make it look untidy and need washing out, though the headmaster did not actually say this.

with their arms round each other's shoulders, laughing, 'He he he' (a female Tweedledum and Tweedledee, and with just as little idea of how to start a conversation). They laughed again, and I laughed, and we were quite content to laugh together about nothing in particular (an enjoyment not comprehensible to adults, who think one must laugh at something and suspect that the target of the laughter is themselves). After sixty seconds' worth of laughing one of them said, '*She*'s got a funny one for you.' 'I haven't,' said the other. '*I* might have,' said the first, and began, 'Fatty and Skinny went for a walk, Fatty blew off and Skinny—I don't know what Skinny did.' The market gardener's son was standing by in his usual paternal role. He looked concerned and said, 'I think they must be thinking of "Fatty and Skinny had a bath, Fatty blew off and Skinny laughed."' 'There's "Fatty and Skinny went to bed, Fatty blew off and Skinny was dead",' said another child helpfully. 'No, it wasn't like that,' said the originator, not much disconcerted and still in the game. 'It was "Fatty and Skinny had a ball, Fatty blew off and Skinny touched it;"' but she retreated with a comic staggering movement as she said it and went to join the girls who were playing Donkey. It is nice to know that Fatty and Skinny are still entertaining children after nearly eighty years.

The market gardener's son looked at me solemnly. 'I'd tell you a story about a pencil—but you might not get the point of it.'

People were running past, chasing. As a boy begins to run away from a chaser his eyes open wider. Further away, by the gate, two girls were dancing an improvised dance. They skipped towards each other, hitting their right hands against each other as they passed, and marked time for a few beats before repeating the process. 'We made it up, didn't we. Well no, we didn't make the song up, it's from Ken Dodd. It's "We are the diddy men, The itsy bitsy diddy men".'[1]

A girl was leaning against the gate post, a thin girl with a red coat and an expression as if she had suffered all the sorrows

[1] Ken Dodd, famous entertainer and comedian, often on television. The 'diddy men' of Knotty Ash are part of his act.

of the world (and how could I tell she had not?). 'You're not playing anything?' I asked, knowing I should not ask, and trying not to sound officious. 'No.' 'Why not?' 'Because I don't want to.' 'Fair enough. Playtime's to do what you want in.' I went to look at the ball games along the big wall: Sevensies, played slowly and carefully (they never were adept at two-balls here), and Donkey. 'Nobody was allowed balls at all, because of the broken window,' they said, as if I ought to have known. 'Most people have got them today. We were allowed them on Monday.' Round the corner the boys were playing football, and the girls netball. When I looked back towards the gate a second girl, propped up facing the first, had joined the silent anti-playtime movement.

I joined the senior girls as they came back balancing their netballs on one hand with jolly public-school nonchalance. 'Any stories?' 'I've got one, but it's rude.' She looked at me with her head on one side, waiting for me to accept her offer. 'There was this Englishman, Scotsman, and Irishman, and they wanted to go to the loo. They came to the public house and the Irishman went in first and there wasn't a light. And he heard this voice,

> If the log rolls over we'll all be drowned,
> If the log rolls over we'll all be drowned,
> If the log rolls over we'll all be drowned.

Then the Scotsman went and he heard this voice,

> If the log rolls over we'll all be drowned . . .

The Englishman went in and he had a torch, and he heard this voice,

> If the log rolls over we'll all be drowned . . .

And he shone the torch down the loo and there were these three little maggots on a—?' She looked at me doubtfully. 'Lump of muck?' 'Yes, that'll do.'

Thursday 16 *March*

Snowing a little; small dry flakes, with the sun shining through them. The 7-year-olds came trooping out of the old Infant

School behind their teacher. A teacher from the main school shouted across the road, and they all went trooping in again.

I wandered forlornly into the playground and stepped inside the school, hardly knowing whether I was hoping to lure some children out or look at the books in the school library. The headmaster came crashing in at the door, a refugee. 'Miserable weather,' he said irritably. 'We've decided to have wet play-time. Awful weather it's been.' I went out again, and watched the young women teachers running daintily through the light snow flakes on their way to have their coffee. Two stalwart schoolgirls came out of the door, each carrying two cardboard cartons of crisps. We walked round to the canteen together. 'I don't know if we'll sell any today,' said one. 'People don't usually think there's crisps when it's wet playtime.'

They took up their stance behind a pale blue formica-topped table in the canteen, and opened the 'till', an Old Holborn tobacco tin with some loose change in it. About ten children came in and bought the 5p packets without, I was told, the usual urgency. All the same a stream of verbal restraint was necessary: 'Wait a minute,' 'What's this for? How many do you want?'; and, when a customer complained about the graphics on her purchase, 'That's the new packet, not the old one.' The sellers were soon left to their accounting. They began counting the number of bags of crisps left. 'Forty-six bags I've got, I must have sold two,' said Joanna, flipping the bags out by their corners on to the table. Then Miss —— spoilt the count by rushing in for a 'chutney flavour', and the extra bag was subtracted aloud, in narrative form. 'Miss —— has bought one, so now I've got forty-five.' The crisps were 'KP Outer Space Pickled Onion Spacestations', and 'KP Outer Space Chutney Flavour Starships', with a Star Wars Fighter Kite Offer, and a round blue disc saying 'Copyright 1977 Twentieth Century Fox Film Corporation'. Joanna watched me writing the particulars in my notebook. 'Your writing's getting better,' she said. 'I can read it this time.'

They went back to their accounting, filling in the columns headed 'Date', 'Start', 'Sold', 'Stock', and 'Cash' in the two small notebooks, one for 'Outer Spacers' and one for 'Outer Spacers Onion'. After much consultation Deborah (the older by a month or two) said, 'Well, that'll have to do. You'll have to tell Mrs —— you've got 5p too much.' Joanna, simply by way of conversation, told me, 'Deborah had too much once, and Mrs —— said if she ever got £1 too much she'd give her a present.' We left aside any comment on the ethics of that. 'Some people only bring £1, and you've got to give them change. Or they bring 50p, or sometimes it's all in halfpence.' 'Yes, and usually they come in all in a rush, and shout.' 'We're only supposed to do it for a week, two of us. This time we've done two weeks. We give up our playtime sort of thing.' I asked, 'Do you mind giving up playtime?' 'Well no,' said Joanna, her round pink practical face very earnest. 'It's cold outside and there's nothing much to do. If we had an adventure playground it would be better. We're not allowed balls again, because Daniel Masters broke a window. The Infant School have got an adventure playground. It's good. Well, it's not an adventure playground really. They've got things painted on the playground and a horse and a climbing frame.' 'Lunch break is too long,' said Deborah with solid bourgeois disapproval. 'You have twenty minutes in the canteen for your lunch and it's a scramble and you have an hour outside and they lock the classrooms so you have to stay outside. If it's wet they don't lock them. We'd better go back now, I suppose.' 'What about these coupons?' said Joanna, 'You get a free packet of crisps if you collect forty-eight—or maybe it's a free boxful. Anyway, come on.'

Monday 20 *March*

A chilly, uncharitable day, a rebuff to spring. I could not see a sign of life in the school; then a reassuringly dwarfish figure appeared round the corner.

Some of the boys had assembled a human horse, and were trying to hoist the too-small rider on to the too-high back of the animal. 'Go on, Midget, I'll give you a budge up. Get on then,' said a boy with the kind of detached humorous face one sometimes meets, with relief, at cocktail parties. He watched as the rider clung to the gyrating horse. 'It's a crazy horsie ride,' he said appreciatively. Then he took me aside (this urbane man) and murmured, 'David's brought his camera—'; but a more ebullient character took over the story: 'Yes, David's nutty. 'E said 'e'd take a picture of people with no clothes on, and 'e hasn't even got a film in 'is camera!'

Neat little girls queued up in new coats (blue or red, with hoods) to tell jokes. 'What is the highest building in a town? A library, because it's got the most stories.' 'What is the quickest way to catch a fish? Get someone to throw it to you.' 'Oh nice,' I said. 'Where did you get them from?' (the automatic question, like 'I like your dress—where did you get it from?'). 'It's a big playbook. There's two pages of jokes.'

The boys wanted to tell jokes, too. But boys do not wait quietly in queues. Boys thrust their faces as near an adult's face as they can reach, and bounce their bodies from side to side, hoping to arrest attention on the in-swing. 'Want any jokes?' they cried like costermongers, bright smiling faces making their wares attractive. I did not really want any more jokes, but found myself saying, 'Just one then,' to a coffee-coloured face framed with dark curly hair. The boy nodded at his friend. '*He's* going to tell it.' (They are both 9 years old.) The friend planted his feet firmly and started the story. 'There's this father and boy and they go out walking. The father 'e bumps into a policeman and 'e goes "Oh fuck!" and the boy says, "Dad, why do you say that?" and the father goes, "It's a word for policeman, son." When they comes to the end of the walk and they gets to the door the father forgets to wipe 'is feet on the mat and 'e goes, "Oh shit!" and the boy says, "Dad, why do you say that?" and 'e says, "It's a word for doormat, son." And then the Dad goes up to have a shower and 'is nose started running and 'e says, "Oh snot!" and the boy says, "Why do you say that, Dad?" and 'e says, "It's a word for shower, son." And the doorbell rang and the boy goes and opens the door and the policeman's there and 'e goes, "Hello fuck, don't forget to wipe your feet on the shit, my Dad's in the snot."' (I must have faltered in my scramble to write the story down, because a helpful small boy leaned forward and spelt some of the words for me—'S-H-I-T', 'S-N-O-T'.) The story-teller relaxed. I could see from his flushed face, and the tall way he stood and looked literally and pleasantly down his nose at the listeners to right and left of him, that he felt he had done a brave and difficult thing. Joke-telling is a useful accomplishment, which may first be learned in the playground.

Perhaps I had been showing signs of restiveness because the next boy was pleading, 'I've got one, it's only a *little* one.' I leant an ear in his direction and caught two little jokes before I moved away: 'What is smaller than an ant's mouth? An ant's teeth,' and 'Why is it so hard when there is a goat behind you? It butts in.' A little girl ran beside me saying, 'Where is it safest to put your money when you go on holiday? In a sandbank,' as I reached the first game of the morning, and that was only a girl walking carefully along a crack.

On the big wall the girls were playing two-balls and Donkey in a lackadaisical fashion, and round the corner the boys were playing football in the traditional football area with whatever vaguely spherical objects they could muster: chiefly rolled-up gloves. A cheering group had just scored a goal and were letting off a stream of insults at the opposing team, one boy ending with the triumphant clincher, 'I'm the football commentator!' Excited spectators at my elbow were dressed in red and white knitted caps and scarves. They were bouncing on their toes and grinning. 'You heard this?' they said.

> One, two, three, four,
> Listen to the Saints roar,
> 'Who's nipped me handbag?'

'There's another one.'

> One, two,
> One, two, three,
> Knickers to the referee!

'That's when they don't like the referee's decision. The Saints? They're Southampton. This is the Southampton colours. Well, it should have black in it really. They wear red and white striped shirts, and black shorts with a stripe down the side.'

A largish girl brought forward a waxen-looking child with black ringlets, and made this statement: 'The boys have been calling her Poofy, 'cos she smells, *and* they say she's got fleas.' Neither of them seemed upset; it was as if they were reporting a leaking tap.

Down the alley-way four girls were trying to lift each other up. Two, holding each other's hands, lifted another two whose similarly held hands crossed their own. 'We can lift them, but they can't lift us.' It occurred to me that experimenting with lifting other people in various ways is a childhood-long occupation.

The bell rang. The bell is a real Lewis Carroll Bellman's bell, brass with a long wooden handle, and a sound that limps, 'De dong, de dong.' After the bell goes the story-telling really starts, for the boys have stopped playing football and there is time to fill before going in to school.

The fourth-years (10 and 11-year-olds) were sizzling with stories. 'There was this man and he wasn't feeling very well so he went to the doctor, and the doctor said, "Do you go to the toilet very often?" and the man said, "Yes, 8.15 every morning." And the doctor said, "You've got three little bugs inside you and I'm going to give you some poison to kill them." And the three little bugs heard this [the tale was told softly, like a fairy tale in the evening] and the first one said "I'm going to hide behind the heart," and the second one said, "I'm going to hide behind the liver," and they looked at the third and said, "What are you going to do?", and the third bug said, "I'm going to take the 8.15 out of here."'

'We've got *lots* more,' they said. 'No more today,' I groaned. 'You've no idea how long it takes to write them out when I get home.' But one pleaded, and lingered with his friend and would not go. 'One joke, *please*.' I nodded. 'I don't know it, '*e* knows it.' He took his friend's head between his hands and whispered in his ear. The friend looked pleased and began reciting,

> Yankee Doodle went to town,
> Riding on a pony,
> He pulled 'er tit,
> She had a shit,
> And then she 'ad a baby.

Friendship is bringing your friend forward to tell his best joke.

Monday 10 *April*

First day of summer term. A brightly coloured sky, with a wind sharp enough to stab anyone's spirit 'broad awake'.

The children's voices as they came out to play were as harsh and invigorating as the wind; perhaps the Almighty tunes them so, to keep parents and teachers in fighting trim.

I was faced by an eager little girl saying, 'I got one! I got one!', as if it were only yesterday that I was collecting jokes in the playground. 'It's Fatty and Skinny,' she said. 'Oh, I think—', I said. 'No, it's a new one. My sister told me in the holidays.

Fatty and Skinny had a bath,
Fatty blew off and made a bubble bath.'

The couplet seemed at variance with the rather governessy navy-blue gabardine raincoat she was wearing, very new and stiffly hooded.

I always feel like a stranger, on the first day of a new term, and I found myself putting a formal, interviewer's question to the crowd of friendly children who had gathered round: 'What does it feel like to be back at school?' They resorted to pantomime, as they usually do. They stuck out their tongues with a faint retching noise, and reeled sideways in a cautious imitation of falling down. When they returned they said soberly, 'After break we've got a maths test. Uck.'

The seniors were walking back from Assembly in a purposeful manner which, even if they not been carrying tubas and euphoniums, would have indicated they were not out at play. They scarcely looked at me. I had to go round to the football area for human comfort, and there, of course, the boys were engrossed in picking sides (although still doing up their trousers after visiting the toilets).

I was adopted by some hospitable riddlers. 'What would you give a sick bird? Tweetment.' 'What would you give a sick pig? Oinkment.' 'What do children in Scotland have for afters? Tartan custard.' 'What do you lay on the road? Tar? You *are* a baby, can't you say, "Thank you"?'

Nothing much was happening in the way of games. Two negligible games of marbles. Girls running along the lines of the netball court. But some girls' games are so quiet one hardly recognizes them as games. I said to a girl who seemed to be standing with nothing to do, 'Not many games this morning, are there?' and she said, 'We're playing one. We're playing Please, Mr Crocodile. Someone's on it, you see, and they say,

> Please, Mr Crocodile, may we cross your golden river
> In a silver boat?
> If not, why not?
> What is your favourite colour?

And then the one who's on it chooses a colour and if they've got that colour on they go over, and if they haven't they have to run and she has to catch someone and that person is Mr Crocodile next time.'

'What other games can I put down?' I asked another girl. 'You see those paving stones over there? Come on, I'll show you.' We went over to the flagged path leading to the classroom block. 'You have to go, "England, Scotland, Ireland, Wales, Inside, Outside, Monkeys' tails." You stand like this and you put your two feet like this over this line, and then over this line, do that twice, then both feet inside the square and then outside the square and then again like at first. You have to do that five times.' 'That's one of the rhymes for elastic skipping.' 'Well, you can use it for that as well.'[1]

A little boy, upholstered like Bibendum, was rolling on the asphalt with a stone in his hand.[2] My first thought was that he was going to throw it at someone. Then I was sure that the stone was what the bric-à-brac shops call 'a collector's piece'. 'Is it a special stone?' I asked him. 'Yes, it comes from the bottom of a bonfire.' The roaring boys arrived and leapt round him yelping, 'Watch it! 'e's dangerous. Small, but violent.' They speak in a kind of primitive rhythm, a gut poetry.

[1] The game on the flagstones was probably derived from elastic skipping, in which a player performs a sequence of cat's-cradle-like movements in an elastic band stretched between the ankles of two 'enders'. Elastic skipping was known as American Skipping when it arrived in this country from America about 1960, and had almost gone out of fashion by 1978.

[2] Bibendum is the fat little man, made entirely of tyres, used as the symbol of the Michelin Tyre Company.

They stood grinning at me like wolves. It would have been fatal to have allowed myself a moment of self-consciousness. 'Well, what's it like to be back at school?' I said heartily. 'I'm finking of leaving,' said the fur-collared cad. 'Anyone know this one?' shouted the fleshy boy with the white eyebrows. 'This man in a cellar, he says, "It's dark down 'ere, isn't it Fred?" and Fred says, "I don't know, I can't see a fing."' 'What about the bloke who found an Irish woodworm in a brick?' 'Go on, tell us the rest.' 'That's all, you dimwit.' Someone on the edge of the crowd said, without hope of being heard, '*The Horse* by Major Bumsore.' The fur-collared cad had already fixed me with his eye: 'There was this lady dancing round the room—starkers.' (A kindly little boy by my side said, 'If you can't spell "starkers" just put "in the nude".') 'She lived up the estate and her name was Collins—she wore ear-rings, actually—um.' He saw my belief was wavering. 'This was on *telly*,' he said, and went on with renewed confidence. 'She said she would give—' Mr ——, on playground duty, loomed over us and said with paternal mildness, 'They've all got peanut brains, this lot.' 'So I've been finding,' I replied.

The bell rang and the boys dispersed. But the fur-collared cad had been following me. 'I want to tell you about some men who went up in an aeroplane.' 'No, not now,' I said, 'You go in and learn how to be a Prime Minister.' 'I don't want to learn that, I want to learn something else.' 'What do you want to learn then?' 'Sex education.' He left me standing, witless.

Friday 21 *April*

Clouds full of rain. A slight but savage wind.

A boy came up to me straight away and said, with the brave swagger that is peculiar to the male sex, 'Are you taking Irish jokes?' That having been settled, his freckled face relaxed and he said, 'Oh good. Well, there was this Irishman, Japanese man, German man, and Englishman and they all wanted to find a hotel to stay the night and they found one and they said to the manager, "Have you got any rooms to spare?" and the manager said, "No, we've only got a spooky one. There is a twelve o'clock ghost in it." So at twelve o'clock there was a bang and a light kept on flashing and 'e went in to the Japanese man and 'e said, "I am the twelve o'clock ghost," and 'e took two minutes to find a knife and stabbed 'is self. So the ghost went on to the German man and 'e took eight minutes to find a pistol and shot 'is self. So the ghost went on to the Englishman and said, "I'm the twelve o'clock ghost," and the Englishman said, "No, you're not, you're ten minutes late."' The boy and I looked at each other with satisfaction. We had accomplished the telling and writing down of a long story. Then I said, 'But where's the Irishman?' He looked at my notebook and round at the rest of the children. Perhaps there never had been an Irishman in the story. 'Never mind,' I said, 'I'll sort it out when I get home.'

Frisbees, descendants of the American Frisbee Baking Company's 'Mother Frisbee' pie plate, have invaded the playground. A stolid bespectacled lad who had been listening to the ghost story was holding one. 'Yes, it's a Frisbee,' he said. 'It comes from the United States of America. You do it like this. You just frow it.' He showed me the correct wrist action, the action one might use at a hoop-la stall. 'I bought this one and then a few other people bought theirs.' He was pushing the centre of the Frisbee (a tough blue plastic pie-plate shaped disc) in and out as he spoke. 'You'll bust it if you do that,' said a friend severely.

Most of the boys had aeroplanes. 'You get them from the toyshop, they're polystyrene and plastic. They cost 30p.' Beautiful little planes of different kinds, to put together oneself. It was the usual tale: 'Someone brought one to school and now it's a craze.' I doubt there will be a plane left in the shop by this time.

'I must go and see what the girls are doing,' I said, and went to where twelve girls were sitting cross-legged in a circle. 'We call it Drop the Glove,' they said, ''cos you drop the glove and you have to run.' A girl on the outside of the circle held up a dark blue nylon glove between two fingers in confirmation. It was good to see that the old game of Drop Handkerchief was still in the repertoire.

In front of the main door some smaller girls came up to

me confidingly. (This is the feminine approach.) 'It was my birthday yesterday,' said a gap-toothed maiden, her harsh little voice not quite up to the feminine ideal. 'I was 8. I got £3 and 43p birthday money. I might have my ears pierced tomorrow, I don't know how much it costs. We're playing Donkey.' 'We *were*,' said her friend. 'Come on, Tracey.'

Further round, the seniors (now frighteningly tall and self-possessed) were practising netball; and netball, like football, is not to be interrupted. I could not see any marbles players, and only one girl had brought a skipping rope to school. I was saved from ostracism by a rush of little girls who wanted to tell jokes. 'Who *didn't* go into the ark in pairs?' they cried, and waited with brilliant eyes while I pretended to try and think of the answer. 'Worms,' they said. 'The worms went in in apples.' 'And who were the last animals to go into the ark? It was the elephants, 'cos they took so long to pack their trunks.' The girls did not leave my side as we walked towards the gate to join the 7-year-olds' convoy. 'I know what you're going to say,' announced one. 'Do you?' I replied. 'No, you must say "What?"' We tried it again, and when I had duly said 'What?' she countered with, 'I *knew* that was what you were going to say.' I felt we could have done with some more practice.

'You'd better come over to the first-year block,' said one of these 7-year-olds, while we waited by the gate, 'everybody knows jokes over there.' So I marched over the road with them, while Mrs —— stood guard against the traffic. The little girl walking next to me looked up and said, 'My Daddy sang a song that I can tell you,' so we settled down (with permission) in a corner of the cloakroom. Other 7-year-olds loitered, saying they knew jokes too, and anyway if they did go back into class they only had 'a little bit of chalking to do'. They tried to beguile me with 'Knock, knocks' ('Boo who?' 'No need to cry, it's only a joke.') but I had been bombarded enough and sent them to their work. 'Tell me about this song,' I said to the child who had been waiting. 'We haven't heard it for a long time, then something reminded him and he taught me,' she said. 'It goes,

> There is a happy land,
> Far far away,
> Where little piggies run,

Three times a day.
Oh, you should see them run,
When they see the butcher come,
Cutting slices off their bum,
Three times a day.

It's good, isn't it? He did sing it to us before, a long time ago. Then he forgot it, and Mummy forgot it, and Jason forgot it—he's 9, he's a right pest sometimes—and I forgot it. Then Daddy remembered it, I think it was because he saw a cartoon about three pigs on the TV, which was on just before a good programme.' I liked her delicacy in explaining how her father came to be watching a cartoon. 'Yes, it is good,' I said. 'Thank you very much.' 'Good' did not seem the right word to describe this gruesome and once very popular hymn parody. Perhaps she liked the tune and the way it all rhymes.

Tuesday 25 *April*

Haze, a strong breeze, the sky greying rapidly, and yet—with dandelions underfoot and swallows above—the certainty that summer is on the way.

A boy stood in the playground, his foot on two footballs, another under his arm; and the keen netball players were already practising. The footballers swarmed out. One of them flipped a green Flippa Disc (from a Shreddies cereal packet), watched it soar over the school fence into the jumble of sheds next door, said, 'I dunno where it's gone. Never mind—come on,' and flung himself into the nearest game of football.

A large amiable chap strolled up, hands in pockets. 'I'm trying to think of a corny joke to tell you,' he said. 'Did you hear about the Irishman who thought the Rover 3000 was a bionic dog?' He beamed at me, luxuriating. 'Oh, and I want to tell you a dirty joke. A boy fell in the mud.' A shorter boy, sandy-haired and rather sharp-seeming (but really as dim as the rest of us) said 'That's from *Our Show*.' I did not catch the name. 'What did you say?' '*Our Show*. It's television.' 'How do you spell it?' It was not until several jokes later that he was sufficiently relaxed to tell me that it is an ITV programme on Saturdays at 9 a.m., and to say the name slowly so that I could understand it; but all the time he had infinite patience. The invincibly amiable boy, meanwhile, had been retailing, 'What do you call an animal that's half a sheep and half a kangaroo? A woolly jumper,' and 'What goes, "Peck, peck, bang!"? A chicken in a minefield. That's from my brother, he makes 'em up.'

Sandra, tall and grown-up, had arrived bearing a Rover biscuit tin. I began to ask her and some other girls about dandelions and other inhabitants of hedges and ditches, hoping to prompt some minor games of country children. Sandra seemed impatient. She metaphorically brushed me aside, and thrust the biscuit tin at her friend Nicky. I thought perhaps she was in charge of distributing refreshments for a school concert. I thought, 'She's too grown-up to ask about games.' Nicky opened the biscuit tin. Inside were half a dozen fat snails, crawling on a bed of grass cuttings and chunks of carrot. Sandra said, 'I don't like opening it, because once they were on the lid and I didn't know where they were. They're so squashy.' She poked one of the snails and said, 'Come on, Fred. That's yours, isn't it Nicky? I've marked mine with purple ink.'

The playground was full of flying objects. Frisbees flew just above the ground (a schoolchild variant, or a prudent instruction from teachers?). Smaller flying discs turned out to be the tops of 1 lb. Blue Band margarine tubs, more manageable, less disaster-prone, free, and just as good flyers in their way. Little girls were operating these, with self-conscious cries of 'Oh! the wind's taken it.' A racketing plastic object I thought might be a boat was 'Not a boat. It's a car that's lost its wheels.' The studious, articulate boy in spectacles took the arm of the wheelless-car owner, put it in a loop against his hip, and aimed a paper dart through the loop. 'What's that called?' I asked, 'an aeroplane?' It was a large dart, folded many times into a Concorde shape. He told me kindly, almost condescendingly, with a sweetness a teacher might find unnatural and irritating, 'It's a sort of aeroplane I've made up myself.' He spoke as if to a child.

An Evel Knievel Fling-a-Ma-Bob (the same size as a Frisbee)

was caught in the top branches of one of the whitebeam trees, and the boys were trying to get it down.[1] They leapt ineffectually at the lower branches, scattering the silver buds, and launched themselves against the rigid trunk. I was annoyed as much by their ineptitude as by the damage to the tree. 'You'd better get the caretaker with a ladder, or you'll hurt the tree,' I said. 'I don't care,' was the reply. 'It's not my tree.'

I went across the front of the school, past bodies writhing on the ground amid the sound of machine guns, and past a game of Blind Man's Buff, the blind man's eyes bandaged with a red, grey, and black school scarf and the two tormentors shouting, 'Stewart! Stewart!' When I looked back at the whitebeam tree only one boy was still assaulting its stout trunk.

The one skipper, bored, handed her skipping rope to a friend. The one game of marbles was not well attended. The netballers stopped their game to say hello, and one of them said at dictation speed, 'As you can see, we are playing netball, and we are winning 6:2. We have won all the matches this term. One match we won 16:2.' The birthday girl from last week came running to say, 'I had my ears pierced. It was £4 and 50p.' I admired the small gold balls she was wearing in her ears.

Some of the children were being allowed longer playtime, it seemed. I wandered amongst them, waiting. The boy who told me the long Irish joke without an Irishman in it came up to me shyly, twitching at the belt of his trousers. 'I got a joke for you.' 'Yes?' In his rich Hampshire accent he said, 'What's better than a butterfloi?' 'I don't know.' 'A centipede with sore legs.' We looked at each other doubtfully. 'I got a rude joke, but I don't know if you want it.' 'Let's try, shall we? They're all jokes.' He began,

> My old man's a dustman, he wears a dustman's hat,
> He took me round the corner, to watch a football match.
> Fatty took a rotten shot and knocked the goalie flat.
> Where was the goalie when the ball was in the net?
> Half way to America like a Jumbo jet.

[1] Evel Knievel was the stunt man who leaped over buses on his motor bicycle.

He had been pleating his grey jersey up his front and pushing it down again while he recited, and looking to left and right. Now he paused. 'Is that all?' I asked. 'No.' He wriggled, and scratched his bottom. 'Um,

> They put him on a stretcher, they put him on a bed,
> They rubbed his belly with a lump of jelly, and this is
> what they said:
> Rule Britannia, three monkeys on a stick,
> One fell off and said, "I—feel—sick."'

'That's lovely,' I said. 'Thank you. But there's one bit in the middle that doesn't seem quite right.' We went through the song again, and still it was soft in the middle. 'Is there anybody else who knows it?' I asked him. He pointed to a boy with an alive triangular face who was playing a chasing game on the netball pitch. 'Kevin knows it, I think.' Luckily Kevin considered it a treat to recite rhymes. He shouted to the other players, 'I'm out of it,' and we went to the wall. 'There's two what I'm going to do,' said Kevin importantly. 'First one coming up.'

> My old man's a dustman, he wears a dustman's hat,
> He wears gorblimey trousers, and he lives in a council flat.
> He took me to a football match, Chelsea was the goal,
> A man jumped up the football post and said, 'Mighty ho!
> Who's lost their knickers?'

'Thank you,' I said. 'And do you know the other one?' 'Oh yes:

> My old man's a dustman, he wears a dustman's hat,
> He took me round the corner to watch a football match.
> Fatty passed to Skinny, Skinny passed it back,
> Fatty took a rotten shot and knocked the goalie flat.
> Where was the goalie when the ball was in the net?
> Half-way up a lamp-post with his trousers round his neck.'

'Half way to America like a Jumbo jet is much better,' said the shy boy. 'I'll put them both down,' I said. The rest of the song was the same. 'Now,' said the bright boy, 'My second one is, How do you make a brass band-stand? You take away their chairs.' 'I don't understand,' said the shy boy. The other explained to him: 'Well, you know people play in brass bands and that? Well, how do you make them stand up? You take their chairs away. I didn't get it myself at first.'

It was time for them to go in. As I went out I nodded to the kindly caretaker, and when I was walking back to my car alongside the playground I realised he had been rescuing Evel Knievel from the tree.

Wednesday 3 May

After the dismally wet May Bank Holiday the weather has repented. The sun is almost too hot, the sky is blue, and the trees, emerging from their screen of mist and rain, are seen to be in summer leaf.

I was buttonholed by Paul's brother as soon as I went into the playground. 'You want to see my lizard?' he said. 'It's in the classroom. Part of natural history, it is.' 'I'd better not now. I'll come in afterwards,' I said, looking longingly at the drinking fountain and knowing I must retain my adult status at all costs. We walked round to the tall wall at the side of the school, which was alive with bouncing tennis balls, yellow ones and white ones. One section had been taken over by the boys, who were dodging back and forth being pelted by a ball. 'Oh, what's this game?' I said, delighted to see a game I remembered from my earliest visits to this playground, years ago, and thought long forgotten (it used to be called Soldiers). 'It's Stinger,' said one of the players. 'No,' said Paul's brother contemptuously, 'it's not Stinger. You play it like—they run across the wall and if the ball hits them they come out here and they take the ball from the person who's hitting them. It's a *kind* of Stinger.'

Paul's brother squinted up at me and took up the confident stance of a salesman offering a good line, his weight on his back leg and his shoulders back, the weight of the shoulders eased by his hands being in his pockets. 'You want a story?' he said. 'Mm,' I said reluctantly. 'About Jack a Nory?' 'Yes, I want it,' I said, getting the message. 'Shall I begin it?' 'Yes.' 'Nuffink in it.' 'Oh I like that,' I said. 'Where did you get it?' 'Dunno,' he shrugged. 'Just made it up.'

'Shall I tell you a joke?' said one of the girls who had been listening. 'No,' said Paul's brother, 'go away.' 'Why shouldn't she?' I asked. ''Cos I don't like her. 'Cos fishes don't talk.' My head swam with all the fish games I had ever heard of, mostly Brownie games of an educational nature. 'I don't understand,' I said. 'Her name's Linda Salmon. We call 'er Fish Face.' Linda Salmon looked pleased, like a kitten being petted. 'What's your joke?' I asked her. 'Have you heard of the three Irish twins?' 'No, I haven't,' said Paul's brother. 'That's *it*,' she said. 'Don't you get it?' 'Oh,' he said, 'I got a better one than that.'

We all moved away from the bouncing balls and Donkey players. 'Once there was this man,' began Paul's brother, his white face heavy with concentration, 'And 'e was walking along the road. This man in a lorry stopped and said, "Where do you want to go to?" and 'e said, "King Street." And this man 'e said, "What's your dog's name?" and 'e said, "It's called a long-tailed, short-footed, long-bodied terrier." ('Just put "terrier", it's quicker,' suggested a tall, rather severe girl with a wire tooth-brace, who had been monitoring the story.) Paul's brother took no notice. 'And this man who'd picked the other one up in 'is lorry said, "I bet you a fiver my dog can beat your dog in a fight in my garden shed."' ('This is a load of rubbish,' said the girl.) 'And when they got back to 'is house, out come a Great Dane, and they put the Great Dane in the shed where the other man's dog was, and they left them there. And one hour later the Great Dane come out nipped to pieces. And then the long-tailed, short-footed, long-bodied terrier come out, and there was nuffink the matter with 'im. And the man who picked up the other man said, "Is that what you really call that dog in Britain?" and the other one says, "No," and 'e says, "It's really called an alligator."' I said to the girl, 'That was rather a good story. What was wrong with it?' and she said, 'Well, it did go on a bit.'

The story really had gone on a bit, and playtime was over. Paul's brother assured me the girls had not been playing any games, 'they've just been talking'; but the girls said they had been playing Horses, and 'Karen's been playing Sevensies—ball-bouncing, you know.' 'What about marbles?' I asked, not having had even a long-distance view of the marbles fence.

'They don't play marbles no more,' said Paul's brother firmly. 'It's not in season.' 'They do now and then,' corrected another boy.

Amid a confusion of people trying to tell jokes ('When is it dangerous to go into the garden? When the flowers are shooting.') and other people teasing an 8-year-old about her love life, Paul's brother said urgently, 'Will you come and look at my lizard now?' He fetched a catering-size Nescafé tin from his classroom and poked about in the grass-cuttings and stones at the bottom. I was afraid for the creature's safety; even more so when one of the boys said, 'It's not a lizard, it's a newt. He found it up the field.' (It was a newt, soft and shrinking under the shifting stones.) 'I got a centipede too—look!' said Paul's brother placidly.

(Punctuating all conversation this morning were complimentary remarks about my pencil, which I decorated with silver sealing-wax when I was 10 years old. Only three inches remain, but 10-year-olds apparently still think silver sealing-wax daubed with red enamel is glamorous and desirable. 'I like your pencil,' they kept saying, 'where did you nick it?')

Tuesday 9 May

A caressing summer's day; a day to envy the swallows their habitancy of the sky.

The caretaker was standing on the pavement enjoying the weather, and I stood beside him. He said, 'It's nice, isn't it? A change from yesterday.' After a while I asked him where he had lived as a boy. 'In the Midlands,' he said, 'Leicestershire.' 'Yes,' he said, 'I went to school in Leicestershire. I don't think we learnt as much as what they do these days. They seem to have every opportunity, these days.'

The playground was slowly filling, behind our backs. No great rush anywhere, just children meandering out of school and letting games happen if they would. Warm May-time maidens with tumbled curls; they clustered round, offering jokes and gossip. 'We're doing a play on Thursday. It's called

Mr Chatterbox. It's a Mr Man story, *you* know.'[1] 'Who's the strongest man in the world?' I made an effort and thought of Desperate Dan. 'No, it's a policeman, because he can hold up several cars with one hand.' 'What do you get if you cross a frog with a can of Coca Cola? Croak-a-Cola.' 'What lies on the bottom of the sea and shivers? A nervous wreck.' (The vocabulary of the jokes is slightly more advanced than their own vocabulary.)

Paul's brother appeared. 'Will you go and buy me a pair of goggles?' he asked me. 'I've got the money. It would only take two minutes.' 'No, I've got to stay here.' 'Oh well, want a joke then?' A girl suddenly grabbed him by the hair, and taking hold of one of the girls in the same way said to Paul's brother, 'Say you love Karen Lindsay.' 'Do you love Trevor Martin?' counter-attacked Paul's brother. 'Aw, he won't play,' said the girl, and let go. 'That's mostly what we've been doing lately. We've been going round getting hold of a boy's hair and a girl's hair and saying, "Do you love him?" and if they don't say, "Yes," you pull their hair till they do.'

'D'you want my joke?' said Paul's brother, 'or d'you want to come and get it?' 'No, you go ahead.' 'Well, there were these three boys called Manners, Mind-your-own-business, and Trouble. Manners went to get some fish and chips, Trouble got into trouble, and Mind-your-own-business went to the police station to find him. The policeman asked, "What's your name?" and Mind-your-own-business said, "Mind-your-own-business." The policeman says, "Where's your manners?" and Mind-your-own-business says, "At the fish-and-chip shop." The policeman says, "Are you looking for trouble?" and Mind-your-own-business he says, "Yes."'

The girls were egging each other on to tell more jokes, enjoying the huddle and encouraging their chosen performers. 'Go on, Marion,' they urged, 'tell the one about toast.' Marion, pink and smiling, drew me into her trap. 'Say "Toast" five times. Yes, now say "Toast" four times. Say it three times. Then two times. Then once. Now, what do you put in your toaster?' I thought. I *really* thought, and I said, 'Toast.' 'It's dreadful,' I said, 'I've been brainwashed.'

[1] The Mr Men books were written by Roger Hargreaves.

'Go on,' they said, 'tell her the one about butter.' 'No,' said Marion, 'I'd better not tell her, she might spread it.' 'There's the wall one,' said another girl, hanging on Marion's arm. 'Oh yes, I can't climb over it, or I'd tell you.' We all agreed that the wall one was not so good.

'What about games? I must get some games,' I said, and went to the strip of playground between the main door and the main gate, which is about as wide as a wide road (fourteen strides wide). In the middle was the first skipping game I had seen in a long while—a tranquil game of 'Apple, apple' engaging eight of the smaller-sized girls. By the wall, a girl holding up her foot and saying, 'I've got red socks,' identified a game of Please, Mr Crocodile; and, further on, children wheeling and squealing like the swallows in the sky were playing Sly Fox, and attaching themselves to the ever-lengthening human chain fastened to the wall. The skippers frowned as they became entangled with the Sly Fox players, and moved their rope to the side. The Sly Fox players became more dominant and excited, and shouted unnecessarily, 'Hey! where's Helen?' and 'Git away, I'm not ready yet!'

On the whole, though, it was a day for less strenuous pastimes. Two girls came and stood in front of me, not saying anything but collecting themselves in the way children do when they are about to give a performance: one can see that social contact has for the moment been broken off while the order of performance is being marshalled in the head. 'Look at my little flea,' said one of them, holding out her right palm. 'He's going to do a somersault in the air and come down on my other hand.' She pretended to watch the little flea doing his somersault, and to catch him on her other palm. 'Now he's going to do two somersaults in the air.' She followed the movements of the flea as he did his two somersaults. 'Now he's going to do three somersaults.' We watched him, mesmerized; but he did not land on the other hand as he should have. Instead his owner darted forward and pretended to pick something out of my hair, looked at it carefully, and exclaimed, 'Oh sorry, wrong one!'

Playtime came to an end in its usual leisurely way. Two girls still sat on the asphalt holding each other's hands and rowing backwards and forwards. One girl had her feet on the other's feet. 'She's meant to fall down backwards and I'm meant to get up, and then the other way round,' said one of them, 'but it never quite works.' Near them two girls were taking turns to bang each other's blown-up cheeks. It was a morning for intimacies.

As we walked to the classroom they begged, 'Can we tell you one last joke? It's about a boy who wanted to go to the toilet, and he held up his hand and the teacher said, "Tell me your alphabet first." So he went "A B C D E F G H I J K."' (The girl stopped and looked fixedly at the ground as she concentrated.) '"L M N O Q R S T U V W X Y Z", and the teacher said, "Where's the P?" and he said "Half-way down my leg."'

The children waiting outside the classroom were amusing themselves with experiments and idle chatter. 'You try banging your ears and saying something—it comes out funny,' one delighted boy advised another. The girls had turned to teasing. 'Linda thinks teddslies are brilliant,' they jeered. 'She calls them "teddslies", she can't say "teddies". She shows off. She brings them to school and keeps them in her desk and cuddles them. Teddies have got long noses. We think they're stupid.' 'They're not stupid, and they haven't got long noses,' said Linda, not caring at all and only answering as a matter of routine.

Tuesday 16 *May*

Definitely summer—could not be more typical.

As I walked towards the playground I tried to analyse the sound of children playing. It chiefly consisted of the piercing squeals which they emit at moments of tension and failure in games; when somebody is about to be caught, or is caught, or when they fail to jump over the ball in a game of Donkey. When I reached the fence a monotonous chanting rose above the general squealing. Two girls were standing in front of a boy who was sitting on the ground; they were chanting, 'Phi-lip, Phi-lip, Phi-lip.' He got up and rushed at them. 'What's he doing?' I asked a girl. 'It's a chasing game, look,' she said. I saw the boy catch one of the two girls at the far side of the playground, and they all came back to the fence to start again.

The playground is an exchange and mart for amusements. Anyone who lacks ideas for things to make or do can see free demonstrations going on all the time: there is always something to copy, something to watch, something to join in—if one is allowed to. A child who comes to school with no traditional lore of her own can go home at the end of the day with a head full of notions.

Even when, as today, there were few well-defined games in progress, most of the children *did* something; they did not sink to the ground in apathetic heaps. The serious and precise boy who will surely be a bank manager one day pursued me across the playground to show me the Robot he had made from a plastic cotton-reel, an elastic band, a matchstick, and a piece of candle. (These gadgets are usually called Tanks, and they crawl along the ground in a tank-like manner, powered by the twisted elastic. The plastic cotton-reels of today are perhaps not so satisfactory as the wooden ones, which could be notched at the edges to give the tanks a grip.) Then the sandy-haired girl came bounding up and told me to 'Think of a number, double it, add ten, halve it, take away the number you first thought of.' Then with a conjuror's flourish she announced, 'The answer is five!' The trick seemed just as magical as when I first heard it nearly fifty years ago.

Another boy volunteered instructions for making a 'throwing arrow'. 'Some people call them French Arrows. You get a hazel stick, and sharpen one end. And you make a slit in the other end, and cut two V-shaped pieces of cardboard from a cereal-packet, and shove them in the slit, and bend them a bit to make four wings, and bind it with some string or wool. Then you get quite a long piece of string or wool—about twice the length of the arrow—with a knot at one end. Put the knot on the arrow, near the flight, and twist the string round once just to hold it—but don't make a knot. The string comes down to near the tip of the arrow, and you hold it while you aim the arrow, and you wind the loose end round your hand. You throw the arrow, and it is the jerk from the end that gives it extra power. Sometimes they go up on roofs.'

I went to see if there were any games. Some boys followed me over to the ball wall, wanting to tell jokes. The standard of two-balls is becoming higher. A girl was juggling two balls against the wall with only one hand, and a half-circle of other girls were watching her like hawks. 'We're trying to see how many times we can do it,' said one of them. A rather villainous-looking boy (yellowish freckles on a ferrety face) piped, 'Why did the baby brick cry?' and went on when I did not answer, 'Because its mother was up the wall and its father was round the bend. It's good, isn't it? What about Raquel Welch? She dropped her handbag in the pub and said, "Fuck me!" and was killed in the rush.'

A smartly turned-out boy waiting his turn appeared to have written his joke down on a piece of paper, gripped between finger and thumb. (Smartly turned-out? I suppose I mean he had his dark hair parted with a visible parting, and brushed to left and right, in the way we used to call 'smart' in the 1930s.) He said to me, 'Which would you rather be, um [he consulted his paper] a train driver, a nurse, a policeman, a model, or an artist?' 'A nurse,' I answered untruthfully and promptly. He entered a mark in the 'Nurse' column. 'What are you doing? What is it for?' I asked. 'It's just for fun, I think. It's me and Andrew and James and Dawn doing it. I just joined in. Dawn might know what it's for.'

On the steps of the temporary classroom an interesting game was in progress. I hesitated to interrupt, but I wanted at least to know if it had a name. The girl who was 'on it' looked at the other girls and said, 'What's it called? Is it Hoppit Lands? Oh, yes,' and was eager to tell me everything else about it. 'We got Land, Sand, and Sea. The top step is Land, the second is Sand, and the third is Sea. Then we got first crab, first fish, first octopus, and first island. Then we got second crab and second everything. And you have to jump from one of the steps to anything that's called out. Like I might say "Land to first island," that's one of the squares on the pavement. We've used up a lot of pavement. We thought of it, and we changed it a bit to make it our own. Well, it was always there, but we couldn't remember all of it. The Land and the Sand and the Sea were there, and the other bit we made up.'

The boys, still lingering, started jumping about and shouting, 'There's a bundle—look! And it's between *girls*!'

'Girls don't often have a bundle?' 'No. They just kick each other and scratch.' 'What's a bundle for? I forget.' 'It's to settle an argument, right? You kick each other on the ground. I have a bundle almost every day.'

Joining a classroom queue I asked a small tough what he had been doing in playtime. 'Running,' he said. 'Relay running.' He jabbed his neighbours in the chest: ''E's fastest, and 'e's second. 'E's only been here 'alf a day, and 'e's second-fastest! We've given 'im a nickname—we've all got nicknames. Mine's Charlie. I'm Charles really. We call Stephen—what do we call you?' 'Beezer,' said Stephen. Charles went on, 'My Dad made up a rhyme about my nickname:

> Charlie, Charlie, chuck, chuck, chuck,
> Went to bed with three young ducks,
> One died,
> Charlie cried,
> Charlie, Charlie, chuck, chuck, chuck.

My Dad comes from Yorkshire. He's in the army.'[1]

I passed another classroom queue on my way to the gate. The girls were standing together, and the boys standing yards away from them. The girls wanted me to write down some evidence of the boys' iniquity. 'Write down, "We don't like boys. Boys have got fleas." Put, "Boys swear and they are all horrible," and "Boys are dumb."' The sweetest of these indignant females said, 'The boys really smell. They stink. They make us feel sick,' and then she looked up at me sideways and said, 'Are you a detective? Mandy wanted to know.'

Thursday 25 *May*

Bright sun and brilliant sky; but the wind is sharp enough to prevent any casting of clouts.

The children were assembling in the old Infants' playground. Most, as they ran singly to join the queue and were exposed to public gaze for a moment, took a comedian's role: they leaned

[1] 'Charlie, Charlie' has been known for at least ninety years.

forward shaking mops of hair, or trudged in slow motion (in disguise, not wanting to appear as their true selves).

When the queue was complete I noticed that every boy, whether slouching or elegantly poised with shoulders back, had his hands in his pockets. Over in the main playground a young woman teacher told me there is no rule against putting hands in pockets. 'If I saw them standing like that in Assembly I might say something, but you see so many men with their hands in their pockets now, don't you.' (The old fear of self-abuse, and the objection that hands in pockets look slovenly and spoil garments, no longer apply. Who, nowadays, would sew up a boy's pockets to prevent such misdemeanours?)

The yellow-freckled ruffian said, 'Look at my fox's tail!' He showed me a strip of reddish fur which might once have graced the collar of a woman's winter coat. 'It's a fox's tail,' he said, 'I took the meat out.' 'How long have you had it?' 'Since I was a baby. Since I was christened. My auntie gave it to me.' 'I had a tail when I was a baby,' he said, prancing around with the strip of fur held in the appropriate place. 'I took it to bed with me. I used to put it inside my 'jamas, like this,' he said, wrapping it round his neck. It was one of the nicest transitional objects I had ever seen.

Two exuberant little girls took his place, wanting to know if I knew why bubble-and-squeak is called bubble-and-squeak. 'It's because it jumps and pops in the frying pan—I just thought of it.' 'Yes, it's fun, isn't it,' I said and it was indeed fun to hear the old joke discovered again.

A gaggle of hilarious boys gathered round, almost drunk with jollity, wanting to know what I had written down. 'It's War this morning,' they said, waving their plastic pistols. One of them was wearing the top part of a camouflaged battledress. 'Put Frilly Pants,' said another. 'That's when you chase 'em.' 'Put Knickers,' said a third. 'Put Fatty Arbuckle,' said the first. 'He doesn't know what it means,' said the second. 'Do you?' I asked. He reeled off in a half-circle and came back, saying nothing. 'Where did you get it from?' I said. 'Out of my mind.' They began jumping around sparring with each other and shouting, 'Fleas! You got fleas! Put down Fleas, Miss.'

A different type of boy had been standing and waiting: the

sort of boy who takes his turn and thinks before he speaks. 'Shall I tell you a rhyme?' 'Yes, please.'

> Said the little red rooster to the little red hen,
> 'You haven't laid an egg since I don't know when.'
> Said the little red hen to the little red rooster,
> 'You don't come round as often as you use'ter.'

'Thank you,' I said, 'that's a clever one,' and he trotted off well satisfied. The little red rooster has amused the middle classes for most of this century, and was a favourite subject for picture postcards in the 1930s.

Some of the older girls were determined to keep me up with my games-reporting. 'Do you know Doctor, doctor?' they said, and immediately agreed on a demonstration: 'Let's show her. Let's get tangled.' The doctor stood aside, and the rest joined hands and turned themselves into a seemingly inextricable knot. Then they called out, 'Doctor, doctor, I'm in trouble!' and she had to come and disentangle them. One of the bystanders said 'That's Chinese Knot,' but the others all knew it as Doctor, doctor.

The bubble-and-squeak pair came running up. 'They're playing Kiss Chase,' they said. 'Kiss Chase is *horrid*. We hate it.' In the angle of the wall by the front door of the school four writhing bodies were pressed, each with the same tousled golden-brown hair. I thought they were girls taking refuge from kiss-chasing boys. 'Are they chasing you?' I asked sympathetically. 'No, we're chasing *him*!' three of them yelled, and withdrew slightly so that I could see their victim pinned into the corner. As I was hanging about with the younger juniors at the gate, the three viragos captured another boy and hung on to his coat as he struggled to get free. The previous victim passed by, passing a languid hand over his brow: 'I've got a headache all over,' he said.

Half-term next week. I must remember not to come.

Wednesday 7 June

Some rain clouds building up. The village was hopeful. 'It's rained everywhere else but here,' they said, 'and the ground is as hard as iron.' 'It's lovely to have some summer, though,' they added, and I thought so too as I passed a family sitting under a red Skol umbrella outside the pub, and saw, in the distance, little girls doing cartwheels in the playground, their skirts flaring like Japanese paper parasols.

By the time I arrived they had stopped cartwheeling and had gone to drink at the fountain. One of them goose-stepped towards me and said, 'Hello. Why did the lobster blush? Because he saw the salad dressing. Lobsters go with salad, you know. I don't know any more jokes, I've told you most of them.'

Poised over one of the slatted drain covers, a boy was Drain Fishing with a bar magnet on a string, watched by a crowd of interested children. 'Just to see what comes up,' he said. 'Would a coin come up?' I asked. 'Only German coins,' he said. 'Oh look, here's a pin!'

Several girls, half a dozen at least, went to their favourite west wall and started doing handstands, their thin little frocks falling over their faces when they upturned, revealing nothing beneath but briefs and tummy buttons. They stayed upside down endlessly, as if to get their midriffs suntanned. A demure 8-year-old who was standing close to me said, 'They're doing lots of handstands on the field, too.' I said, 'Why don't the boys do handstands? It's always the girls. Are the girls better at them than the boys?' 'My brother can do them better than me,' she said. 'They don't *like* doing them so much. They just like watching the girls' knickers. My brother does. And they don't do them in case the girls say, "Your flies are undone." The girls say that to make them get down from their handstand. They say, "Your flies are undone," and the boys say, "Go and get a camera." Well, the boys can't really do handstands, anyway, and they don't like doing them. They like fighting better, and rough things like that.'

A long game of Piggy in the Middle was in progress right across the front of the playground. 'We've got two in the middle,' they shouted, and I could see there were two or three throwers at each end. In and out of the game staggered a pair of girls 'practising three-legged for Sports Day', their ankles shackled with a red elastic belt. The Sly Fox players were too busy to talk to me, and so were the Donkey players. Some idlers gathered round and said they had been playing Mother, May I? and Cat and Mouse, and Poor Puss, and netball, and Chase, and 'the boys are still playing football but nobody is playing marbles', and Danger Letters. 'Danger Letters? One person is on it and says, "The Danger Letter is A," and then she says something like, "D, take four steps," and if there is a D in your name you take four steps, but if she says A and you take any steps you have to go back to the beginning. It makes you think.'

'What about games with words?' 'Oh yes, we've been playing a good one up the field. You say, "Do, dare, double dare, love, kiss, or promise," and someone's got to choose one of them, and if they say, "Kiss," they've got to go and kiss someone, and it goes on like that.'

The bell rang. The footballers gradually stopped playing and went and stood disconsolately near their classrooms. 'Nothing to do now,' said a freckled beanpole. 'Just got to stand still.' A boy showed me a white plastic monster pen. 'Look at this. Know what this is for? It's for four things. First of all it's a pen. Then it's a whistle. Then it's a ruler. Then it's for hitting Mark over the head—clonk!' The riddlers took over. 'What's yellow, smooth, and deadly? A shark-infested custard.' 'Why did the apple turnover? Because it saw the banana split.' 'Why is it dangerous to rob a bank? Because there are so many coppers about.' (They agreed that no one is cheeky about policemen these days. 'We only call them coppers. We don't tease any grown-ups, only our mothers.') On this not-even-very-hot day they were as slow to understand each other's jokes as I usually am. Thirty seconds afterwards they were still groaning, '*Oh—oh*—I get.' I still find myself falling into an old trap and thinking that if one child knows a joke all the others in the school must know it too. 'Knock, knock,' they went on, 'Who's there? Isabel. Isabel who? Isabel necessary on a bicycle?' 'That's a good one,' I said. I did not add that I knew it when I was 7, that it was the first 'Knock, knock' I ever heard, and that it seemed miraculous because I had a friend called Isabel.

Wednesday 21 *June*

Sunny, after a long shower. A restless wind.

The twenty-two children of Miss ——'s class, lining up in the old Infants' playground, seemed reluctant to face the elements. The girls were hugging their bodies like timid bathers, and seemed glad to be given the word to move across the road and into the main playground.

Three of the older girls were already absorbed in a pretending game. The largest hopped on her hunkers after a mother figure, who asked, 'What do you want? A bar of chocolate?' As I lingered near them I heard the mother (the mother is always the organiser) directing where the others had to sit on the concrete drain cover. 'We'll do that thing we did the other day. You get here at the back and keep your feet together, and I sit here.' When they were settled I ventured to ask them the name of the game. 'We're playing Mummies and Daddies.' 'What's this then?' I said, pointing to the concrete square. 'It's our caravan.'

I said to two other girls, 'Let's go and see what they're doing round at the side.' 'They've been playing Donkey, and Sevensies,' they said; but the only person we saw was upside down, doing a handstand with a crisp bag in her mouth. 'What's this game, I wonder?' I said as we approached three girls kneeling in a triangle. One of the players looked over her shoulder and said, 'It's called One Knee.' The girl by my side said mildly, 'I call it Butter Fingers.' 'I call it Butter Fingers, too,' said the other. 'You throw the ball to each other and if you don't catch it you go down on one knee, and then two knees, and then sit down, and after that you're out. I really think Butter Fingers is a better name.' They were not pressing the point; they were feminine girls, with conciliatory natures.

These gentle children, palely freckled and fringed, accompanied me further. We talked about the absence of Jacks and marbles and hopscotch. 'I don't think we're allowed to chalk hopscotches any more,' said one meekly. 'Well, I asked the headmaster the other day and he said there was no rule against it,' I observed. 'Oh. Perhaps I'll bring some chalk then.'

The other said, 'I remember when Mr —— was here (he was deputy head) this boy kept drawing love hearts all over the playground and writing his girl friend's name in them and Mr —— washed his mouth out with soap and water.' 'That seems a funny thing to do,' I said, and I meant the mouth-washing, which was the Victorian punishment for foul language. 'He used to often do that when people were naughty. He did it when a boy put some itching powder down a girl's back.'

The summer encampments have begun. Susan and Pauline—9-year-olds—were sitting primly at the top of the classroom steps, with their legs straight out before them and their sewing things in biscuit tins beside them, sewing away at some work of their own. Another girl said, 'Guess what I've got in here,' and held up a small brown-paper bag. It felt squashy. She opened it so that I could look inside, at yards and yards of green and red woollen chain. 'It's finger-knitting,' she said. 'I made this, look!' and she pointed to her belt. 'Me and my friend are doing it. A girl taught me.' ('And me,' said the friend.) 'And we copied her and we done it by heart, didn't we Tina?' They showed me how to do it, so that I too could get it by heart. 'You make a loop on your finger [the first finger of the left hand] and another loop next to it, and you put the second loop over the first and you've made a chain.' I think I shall be able to do it, but mine will be a much coarser chain than theirs because my finger is many sizes bigger.

After the bell rang I asked one of the footballers if he knew the song about the Scottish team in the World Cup. He knew it all right.

> We're with Ally's tartan army,
> We're going to the Argentine,
> And we'll really show 'em up
> When we win the World Cup
> 'Cos Scotland are the greatest football team.

'I sing, "We're the worst football team," now,' said another boy. But they none of them knew the disenchanted version that boys were singing down in Lilliput, Dorset, on 14 June:

> We are all Ally's fairies,
> We've been to the Argentine.

We've had a swim and bath,
And played football—what a laugh!
For Scotland are the worst football team.

Scotland were knocked out of the Cup on 11 June.

Tuesday 27 June

A cold, disagreeable day. Oberon and Titania have not yet made up their quarrel. I left my visit till the dinner hour; it had been drizzling during the morning. The children told me they been out at playtime, though, and when I met Mr—he said, 'We don't keep them in unless it's a monsoon.'

The playground seemed forlorn; half the school was away on summer expeditions to London and Brighton. However, as soon as I was asked, 'What do you do when you are hungry?' and was told straight away (in the usual kindly fashion) that I should 'Run round till you get fed up,' I knew that everything was all right. The riddles came thick and fast. 'What is black and white and red all over?' said one, and flattened my confident answer, 'A newspaper' (for I knew that riddle as a child) with 'A zebra with a red cloth on its back.' 'A zebra with jam on—that's what *I* say,' said another. Zebras led to other zebras. 'Did you hear about the Irishman who called his zebra "Spots"?' 'What do you get if you cross the zebra with a pig? Striped sausages.' The chief comedian was a vivacious blonde with a beautiful eggshell complexion. 'I'm the one who tells you jokes—ta ra!' she said, striking an attitude. 'Here's another one. What's the most unfortunate letter in the alphabet? Will you write it down, please. I like you writing them down.' 'Yes,' I said. 'Now tell me the answer.' 'The letter U—because whenever there's trouble you always find U in the middle of it.'

'I really ought to be finding some games,' I said, looking at some little girls going by holding bright new yellow tennis balls. 'They're only going to play Donkey,' said the blonde dismissively, 'and there's some people playing marbles, too. I'll tell you a game. It's a way to keep warm, really. You spin round like this, then you kneel down and kind of bump yourself, then you jump round and when you're worn out you sit down and think how warm you are.' She did not want to leave the stage, but she had run out of material. 'Now I'm going off to play,' she said. 'Oh, *what?*' I pleaded. 'A Kings and Queens game. I'm the King. We've got two Julies as maids. They clean the house and do things like that. And we've got a Lisa and a Stephen as slaves, and Eunice is the Queen.' 'Oh dear, have you kept them waiting?' I said, conscience-stricken. 'We've been playing it for ages. I suppose I'd better go off now—they'll be starting to get a bit worried.' Of such nonchalant stuff are Kings made.

I walked around on my own. In the alley-way between the toilets and the classroom—a haven for fragile games and quiet people—a small boy in a grey cowled duffle coat was sitting on the ledge reading *Alice's Adventures in Wonderland*, sliding a strip of white cardboard down the page, line by line. Some girls, standing at the corner out of the way of the football, were more inclined to talk. 'What did the black man say to the white man?' 'I don't know.' I was feeling lethargic. She sang: 'Cadbury's took me and they covered me in chocolate.' Another girl sang it, and another. 'Does it come off television?' 'No, it doesn't come off television.' 'Who do you think thought of that?' 'Don't know. Probably somebody who was eating Cadbury's chocolate.' It was a soggy conversation. 'Shall we show her this one?' said the smallest. They straightened themselves and began clapping idly:

> I went to a Chinese restaurant
> To buy a loaf of bread, bread, bread,
> They wrapped it up in a five-pound note
> And this is what they said, said, said:
> Ennamy jennamy,
> Om pom poodle,
> Wally wally whiskey,
> Chinese chopsticks,
> Indian chief say 'How—bow, wow!'
> Boom, boom![1]

'I do this when I say, "Boom, boom!",' said one of them (it was

[1] See I. and P. Opie, *The Singing Game* (1985), 465–7.

the expert handstander, now that I looked closely) and pulled out her jersey over the two places where her breasts would be one day. Conversation flagged again. 'Why do you do this writing down?' asked a thin little girl with huge brown eyes. 'Oh, because—it might come in handy. I always like games. And nursery rhymes.' The two hand-clappers started to sing, 'The bear went over the mountain.' 'That's a nursery rhyme,' they said, and they may be right. 'We'll do "Popeye the sailors' man",' they volunteered, and began clapping, but before the second line they had to rush off for their dinner. 'Come on, Janet, we'll do it,' said another girl, taking over as hostess:

> I'm Popeye the sailors' man, full stop;
> Lived in a caravan, full stop;
> He opened the door,
> And fell through the floor,
> I'm Popeye the sailors' man—
> Comma comma,
> Dash dash,
> Full stop.[1]

And who would worry about the exact wording, as long as the hands were placed on the hips and the hips wiggled at 'full stop', the right fist waved in the air at 'comma, comma', the skirts flicked up at 'dash, dash', and the hips wiggled again at the finish. 'I wonder if I've heard that here, before,' I murmured. 'We used to say it and then nobody said it any more,' they explained.

Laura fetched me to see some baby ducklings 'who have lost their Mummy' in the long grass the other side of the school entrance. We could not see them, but we heard them before we were wordlessly shooed back into the playground by the dinner lady. I thought, anyway, it was time for me to go home (time stretches during the dinner hour). I picked my way carefully through a game of Mother, May I? in the alley, and as I emerged into the main playground I was told, 'A boy over there wants you.' It was Paul's brother. He left a marbles game and said, 'I want to tell you a story. Do you know the one about Germany and France?' 'Let's sit down,' I said, and he began.

[1] See I. and P. Opie, *The Singing Game*, 471–2.

'There was this man went up in an aeroplane and he was going over Germany and he said to the pilot, "I want to have a shit," and the pilot said, "Open the door and go out there." Then they was flying over France and the man said, "I want to have a slash"—that's go to the toilet—and so 'e did like before.' He suddenly came out of his dream-like story-telling state and yelled to Paul, ''E's goin' to take our marbles. Go an' get 'im!' A second or two later a tall boy—a stranger—came over, holding a long brown cigarette tin, and said soberly to Paul's brother, 'He only took two marbles to play with.' 'Oh. OK,' said Paul's brother, and turned back to me. 'Just a minute,' I said, because I was writing down the interlude. 'That's not in the story,' said Paul's brother severely. 'I know,' I said. 'Now you can go on.' 'So they went back to England, and the general said to the pilot, "Is there any news from Germany and France?" And the pilot said, "Yes. Germany has just been bombed and France has just been flooded".' He relaxed as he came to the end of the story. 'Thank you,' I said, 'I haven't got that one. Where did you learn it?' 'I got it from my brother Paul.' 'Where did you get it from, Paul?' he shouted. 'Made it up,' shouted Paul. 'You *didn't*.' They went back to their marbles, and a boy leaning against the fence near me said, 'I've got a story. You want it?' 'Is it long? Is it a verse?' 'Yes, it's—

Smack your chops,
Lick your dicks,
Eat a lovely plate of—
Fried shit.

I made it up on an advert. It's the one about Chipsticks—long crispy things, you know, and they're yellow. There's another one,

There was a man from Brazil,
Who ate a dynamite pill.
His heart retired,
His bum backfired,
And his willy shot over the hill.'

I thanked him and said I really must go home now.

An old acquaintance met me at the gate. He held out a

handful of marbles and said, 'Look at these!' 'I love those chinas,' I said, 'aren't they pretty? I keep meaning to buy some.' 'I'll win you one,' he said, and flew off to the marbles fence. 'The trouble is,' I said to nobody in particular, 'that I won't be here long.' As I reached the other side of the road I heard a loud 'Yoo-hoo-ing' and there he was waving his arms about and shouting. When I went back he presented me with a china, streaked with red and yellow and green. 'I just picked it out of the hole, like that,' he said.

Monday 3 July

Black wind-blown clouds; but fine.

As I walked up the road I could see children jumping off the classroom steps, squealing as they jumped and putting their hands over their mouths when they jumped badly. Other children were leap-frogging. Both games had stopped by the time I arrived.

A skipping rope tied on to the far fence (the marbles fence) encompassed one skipper. Another rope with one skipper and two spectators, in front of the main door, made me think a skipping season had started; but when a house-martin fell from its nest a few yards away the skipping stopped and never started again. A crowd of distressed little girls asked me if I was tall enough to put the martin back in its nest. I hardened my heart and suggested a teacher. A phrase from this morning's wildlife broadcast about fledgelings drifted through my mind: 'rescued by well-meaning people.' The child holding the house-martin stroked its head consolingly (do birds like having their heads stroked?). Later I was told that Tracey had put it back: 'She had a piggy-back and put it back.' She must have put it back in the guttering, above the place where it was found. The nest, I saw, was round the corner under the eaves; but the bird was, in any case, only interested in leaving home.

I found the girl who had been skipping. 'We were skipping "On the mountain stands a lady",' she said. 'You sing,

On the mountain stands a lady,
Who she is I do not know,
All she wants is gold and silver,
All she wants is a nice young man.

Then they pick a boy, and they sing the name. They picked Gary for me. They sing,

Now then whoever's-in-the-middle,
Don't tell lies,
We saw you kissing somebody-or-other,
Round the corner.

Then it goes,

How many kisses did you give him?'

(A boy looked over my notebook and, seeing an incomprehensible scrawl, said, 'You write like a Martian. Let me look and see if your hair is green.') The girl continued, 'Then you see how many you can skip, and then they say, "True, false, true, false," until you land on the rope, and if you land on "True" it's true, and if you land on "False" it's false.'

The boy had been lingering nearby with a friend. 'You heard this?' he said. 'It's about a hundred nuns and a sister, or something like that. This sister said, "One of you nuns had sex last night. Ninety-nine nuns said, "Oh!" [in a shocked voice] and one nun goes, "Oooh!"' 'We found it in a magazine, didn't we,' he said to his confidant. 'Probably she was pregnant.' 'You girls should go away,' he said suddenly, and gave a shove to some girls who had been eavesdropping.

I wandered away with them. They were 7- and 8-year-olds, and rather at a loss for conversation. 'Suppose you wanted to play a chasing game,' I asked, 'How would you decide who would do the chasing?' Without the slightest hesitation one of them said, 'We dip. "Dip, dip, dip, My little ship, Sailing on the water like a cup and saucer, You're not It."' 'Or we do Spuds—you know, "One potato, two potato, three potato, four,"' said another. 'Here in the playground?' 'Oh, yes.' And yet I have not noticed any dipping here for at least six months.

We walked sedately round to the big wall, and gazed at a

chubby figure propped upside down there, her head on a folded red anorak, her panties decorated with a marching drummer boy, and no vest. 'Ooh, I hurt my head,' she said as she came down.

The girls became quite chatty. 'My friend told me ever such a funny rhyme,' said one of them. 'It goes,

> Ta ra ra boom de ay,
> My knickers flew away,
> They had a holiday,
> They came back yesterday.'

Her neighbour began to tell me about a hole she had had in the elbow of her jersey, and then said, 'Oh!' and ran off to join a game of Sly Fox. The small roly-poly lisper sidled up to me. She was holding two crisps on her open hand. 'Have a crithp?' she said. I chose one. 'I picked them off the ground! I tricked you!' she said, laughing richly, and I laughed with her. All the while odd scraps of playground conversation had been drifting by: 'My baby's a toddler, and he's all over the place making messes;' 'She says her pink mouse is bionic—it's not even a real mouse.' It is difficult to separate reality from fiction in the playground. The two are in a happy state of confusion; like dinner-party hilarity, when nonsense rises on bubbles of champagne and must not be foundered by leaden fact.

Raconteurs do not willingly relinquish their victims. The boy had been lurking in the background and now came confidently forward with another story. 'There's this man who had this flat nose. It didn't never grow. He didn't like it, and he went to the doctor and said, "Doctor, my nose won't grow." And the doctor said, "Drink this, and every time you move it will grow an inch." So he went out and he was walking along the pavement and he bumped into a lady and his nose grew an inch. And then he bumped into a boy and his nose grew an inch. And then he bumped into a Pakistan lady and she said, "A thousand pardons!" and his nose grew a thousand inches!' He looked at me triumphantly, waiting for this oriental wonder to sink in.

Two boys shuffling their feet near us were undecided whether or not to tell me what they had been playing. One said doubtfully, 'It's not true.' The other said, 'I was a cowboy and 'e was a cowboy too and there was a sheriff and there was a robber who robbed trains and all things like that. We was 'aving a fight and I takes 'im to prison and 'e escapes and that's all.' The bell rang and as they went off they called out cheerfully, 'Part Two tomorrow!'

Behind them two girls had been doing some arm work that I did not recognize. Not hand-clapping, I had thought subconsciously. They were thrilled with the game, obviously, going on and on with it, and laughing. 'I call it One, Two, Three,' said one of them. 'That's paper, when you hold your hand flat. And that's rock, when you clench your fist. And that's scissors, when you hold your fingers like a pair of scissors. Paper wins over rock, rock wins over scissors, and scissors wins over paper.' 'She just taught it to me,' said the other girl. 'Oh,' I said, knowing I had never seen the game before in this playground, 'have you come from somewhere else? Where did you learn it?' 'From my cousin. I'm not *new*.' 'Where does your cousin live?' 'Catesfield, near Titchfield. I met her last year, I should think, and I only just remembered it.'

A boy opened his hand for a moment and gave me a tantalizing glimpse of marbles as he went by on his way to class. 'Won 'em off Terry Masters,' he said, 'Fourteen chinas and a bull-bearian.' '*Oh*,' I cried, and followed him up the steps, yearning for the shining marbles, like the elf in the poem who cries, 'Give me your beads!' The headmaster blocked the way, talking with a teacher. 'Mr ——, may I borrow this boy for a minute?' I implored. 'Borrow him for a month if you want,' said Mr ——, and we retreated down the steps. 'Who's been playing marbles this morning?' I asked him. 'Just me and Mark.' 'How did you get the idea?' 'I just saw Michael playing last year and I still remember after all that time,' he said proudly. He was conscious of giving a good interview. We gloated over the chinas, milky glass with streaks of red, or blue, or green. One was a china-fourer. 'And this bull-bearian, this isn't the smallest bull-bearian you can get. You find them in bicycles as small as *that* [he showed a tiny space between his finger and thumb]. I got a lot at home. I went down to WRP Motors, where my Dad works, and saw a drawer marked "Ball Bearings", and I found an empty cigarette packet and put them in, 'cos Dad said I could 'cos he was deputy manager there.' We parted with a mutual sense of fulfilment.

Monday 10 *July*

Camera-shutter weather, black clouds alternating with bright sun.

A hurried visit, this week; there was too much happening at home. I was no sooner inside the gate than someone was leaping up and down by my side and saying, 'Come quick! The water's spitting up in the air.' She ran back to join the crowd by the drinking fountain. 'It doesn't usually go like this,' she said, as the boys pressed the lever and made the water squirt four feet in the air. They were putting their mouths over the nozzle and jumping back, innocently surprised, as the water soaked them and the bystanders. They jostled for turns, and a circus line developed as they squirted, were soaked, were pushed on, and ran round for another turn. It was like Trafalgar Square on Jubilee Night. Mrs —— came out to quell the excitement. 'Stay away from there, please,' she said decisively. I was glad of the special dispensation which exempts me from behaving like an adult. I wondered if she thought I should have stopped them from frolicking in the fountain, and knew that once I began wondering such things I would be lost. Instead, I attended to a patient little girl who had been wanting me to write down her rhyme:

> Ask your mother for sixpence
> To see the new giraffe.
> It's got pimples on its whiskers
> And pimples on its—
> Ask your mother for sixpence
> To see the new giraffe . . .

'And you go on like that. Is that shorthand you're writing?' 'It's a kind of home-made shorthand.' 'Have you heard about the Irish wolfhound? Well, 'e was chewing this bone, and when 'e stood up 'e only had three legs.'

We were interrupted by two little girls flying past and screaming: 'Miss, stop her! She's hitting me!' 'No I'm not, I'm playing a game.' 'We're not playing a game.' 'Yes, we are, we're playing Chasing.' 'No, you're not, you're fighting, and I don't want to.' They whirled on across the playground, still scuffling.

I had no chance to move to another part of the playground, for the chief story-teller and several hangers-on were standing by my side, waiting. The story-teller likes to be brought forward; I have never known him to approach me directly. Paul (who is often master of ceremonies in the playground) said, 'My friend has got a joke.' 'It's about three fleas,' the story-teller began. 'Two fleas went straight up on to a mountain and the other went down to the woods—the dark forest—and he found a cave in the middle and this big bad hairy long juicy thing came sliding along and went into the cave.' ('Ugh,' said one of the listeners.) 'So the flea went back to the other two fleas and said, "Have a guess what I've seen." "What. A ghost?" "No, I've seen a big bad hairy long juicy thing." ('Poor old flea,' said the listener.) 'And the other two fleas said, "Well, when you go down there take a bucket of water and throw it over the big bad hairy long juicy thing—"' (The audience were in perpetual motion, corkscrewing round the teller, bouncing their tennis balls abstractedly, blowing rude noises on to their forearms.) 'And so he said, "OK," and so he went and got a bucket of water and went down to the dark forest again, and when the big bad hairy long juicy thing came he threw a bucket of water over it, and he run back to the other two fleas and they said, "What happened when you threw the bucket of water over it?" and he said, "Nothing." "Didn't it even get wet?" and 'e goes, "No," and they said, "Why not?" and 'e said, "'Cos it had a raincoat on."'

After that I had to leave.

Tuesday 18 *July*

Half-heartedly summery; light cloud, light breeze. School breaks up on Friday.

The school had an unwonted air of formality. Children were being marched about in lines. Even the 9-year-olds came out of their classroom in a line, although no teacher was to be seen. They halted uncertainly in the middle of the playground. I went to see what it was all about. 'It's the new people from the Infant School,' they said. 'They've come to see what it's like.' Then they chanted rhythmically: 'The first-years have gone up to the second-years, the second-years have gone up to the third-years, the third-years have gone up to the fourth-years, and it's happening for two days.' No wonder they had come out so cautiously, they had been in a strange classroom.

They moved forward to see the new children being brought into the playground. 'Oh don't tell me we're getting Kevin Clegg!' said one of them disgustedly. 'He's horrible. He calls me names. He lives down my road.' The flat-faced girl folded her arms and began walking about and inspecting the newcomers, her head tilted back. 'Are you supposed to be doing something about them?' I asked. 'Yeh,' she said, making a wry face, 'looking after them.'

The older inhabitants were definitely put out, and irritable. The boys pushed the girls, and the girls said, 'Ow! Shurrup!' in snarly voices. I had to go to the western part of the playground to find peace and stability. There the boys had settled on the steps to play Trumps with car picture cards. I asked a spectator how it was played. 'You get the cards and they've got numbers on and if a person plays a higher number he wins.' Each card has a picture of a car on the front, and its specifications on the back. With much questioning I found out that they agreed beforehand which of the many figures was to be used. It might be the maximum speed, or the capacity of the petrol tank. You can buy the cards anywhere, they said. In Woolworths, or in the village.

I could not see any games in this whole large area. 'What about skipping?' I asked a girl. 'Yes, yesterday, that's all,' she said, in the tone of one whose pattern of life has been disrupted. 'We were Building Up Bricks,' she said. A few people were flinging themselves casually into handstands. A game of Sly Fox was being organized, and soon the girl who was 'on' was standing in the characteristic S-shape facing the wall, and bobbing her head back and forth as she called out 'Saw Marion. Saw Carol. Saw Lisa—come on Chubby!' Those girls came up to join the growing chain.

A boy with a long, adult face and a cotton hat asked me, 'You want a story?', rather as if he were offering me a toffee. 'Well yes, thanks,' I replied. 'There was an Englishman, a Scotsman, an Irishman, and a Welshman in an aeroplane—' A girl took him by the collar and said fiercely, 'The Irishman does it best, right? My Dad's an Irishman.' The boy rearranged the story in his head and proceeded. 'Well anyway, they only had three parachutes. The plane got into trouble and the pilot called, "Bale out! bale out!" The Scotsman said, "I'll have to go first, 'cos I'm the brains of Scotland," so he took a parachute and jumped. The Irishman said, "I'll have to go next, 'cos I'm the brains of Ireland", so he went. So there was the Englishman and the Welshman left, with only one parachute between them. "You have it," said the Englishman to the Welshman. "We can each have one," said the Welshman. "The Scotsman took my haversack." '

The retailer of rather poor stories had been listening, and wanted to tell me a joke: 'What is yellow and swings in a tree?' 'I don't know,' I said obligingly. 'Marzipan,' he said. 'I don't get it,' I said. 'Nor do I,' said the market gardener's son, who had been waiting patiently holding a flat wooden box. The dimwit retreated in confusion a few paces and said, '*I* don't know. Somebody told me.' The market gardener's son reckoned the answer must be 'Tarzipan'. 'Oh yes,' said the teller, 'that's right.'

A more accomplished raconteur pushed forward, a burly chap who said, 'There was this lady was going on this truck and this man said, "D'you want a kiss and a cuddle under the truck in a lay-by?" and they stopped in a lay-by and then three minutes later a Panda come along [he paused to collect his thoughts] and a policeman got out and said, "Hello, 'ello, 'ello, what are you doing?" and the man said, "I'm adjusting the clutch", and the policeman said, "You'd better do your brakes 'cos your lorry is five miles down the road." ' The audience stood in a semi-circle, letting the story sink in and murmuring appreciation. 'I got that off a film,' said the storyteller, pleased. 'TV, do you think?' I said to the market gardener's son. I always consult him on questions of fact. 'Yes, I should think so,' he said. 'What's that box you are holding?' I asked. 'It's the Club. I'll show you.'

He opened the large flat box and took out a T-shirt which had painted on it in green 'THE TROUBLE MAKERS' CLUB'. The box also contained chalks and paints. 'Gary Watson started the club,' he said. 'We've got lots of people—at least above twenty. Coming up to forty, I should think. Gary said we were to wear these things when we come to meetings.' The dim boy leaned forward. '*When* do we wear them?' he asked. 'When you come to the club,' said the market gardener's son. 'Are you a member?' I asked the dim boy. 'Yes,' he said, 'I'm the Deputy.'

I was introduced to Gary, the club president. He was sitting on a dustbin near the classroom steps. Other boys, hearing me interviewing Gary, became restive and said they had a club too, established four years ago, 'with a secret club ground up the field that no one knows about.' 'Soon it'll be public 'cos—' blurted out one boy, and was shushed into silence. 'The girls keep trying to find out where it is,' they exulted.

Thursday 7 September

Term began on Tuesday. Warm wet westerly weather; it had only just stopped drizzling.

For a moment I panicked, felt the playground to be a strange place, and could not remember anything about children or their games. Nothing in particular seemed to be happening, only young teachers running across the playground to the main building for their morning coffee. Then I saw two little girls struggling to get their hands clasped behind their backs. 'What do you say, when you play that game, I forget?' I asked them. 'You say "Teddy and me went out to tea, Lock the door and turn the key," and you go marching off across the playground and when the words are finished you turn round and go the other way and *your hands don't come undone!*' It is true, it is quite magical how the hands do not have to be unclasped.

Some boys ran by. One of them looked over his shoulder and shouted, 'No taking over!' 'They really mean "No overtaking",' said a senior girl. 'That's what they say when they're running round the school building, seeing how many laps they can do. They make a lot of fuss about it, who can do most laps. Then they say, "Oof! I've been round *ten* times!" and stagger around as if they're going to die.'

A moon-faced girl, of the wholesome extrovert type, said pleasantly, 'Can I do a joke on you?' She drew circles on my back, saying, 'I make a hole in your back.' There was a pause. 'I threw it away,' she said. Then, 'I put a piece of string through it.' Then, 'I'm pulling you, I'm pulling you.' I found myself obeying, and falling backwards—but not without considering the consequences. 'You're supposed to fall backwards and I catch you,' she said, and performed the rite on a succession of lightweight maidens who fell into her arms with abandon.

At the edge of the playground a new boy in a bulky red anorak which almost hid his grey shorts was allowing his sister to re-tie his shoe-laces. His shoes were shiny new, and had shiny nylon laces. He was completely incapacitated, weighed down, and immobilized.

Suddenly a large, diffuse chasing game was all around me; it was only to be defined by the figures facing inwards, jeering at the chaser, and yelling, 'I'm free, I'm free!' A dancing figure at the edge called out, 'I've got "Scribs!"' and held her crossed fingers in the air (though what excuse she had to claim immunity with the truce term I could not see—usually the children can only opt out if they want to go to the toilet, or if a shoe has come off, or there is some other reason why it would not be fair if they were caught).

The boys at the fence were waving a black, too-realistic pistol. One of them was crouching and grubbing in the loose tarmac; he seemed too clean a boy, with neatly pulled-up socks, to be poking in the dirt. 'What are you doing?' I said in my mildest voice. His head came up and he looked with glazed eyes into the distance. Perhaps he wondered if he was going to get into trouble. A boy who already knew me leaned over and said, 'He's picking up stones to fire out of the pea-shooter—that pistol pea-shooter we've got.' 'Do you have to have stones?'

'No,' said the boy doing the stone-picking. 'You can use peas, or you can get little fish weights in packets.' 'Peas are too big,' said the other boy. 'You have to cut them in half.'

I was being shown a miniature super-ball[1] when the small bouncy batty boy arrived (either he is daft or he is a genius). He has got gorillas on the brain. 'Do you know about gorillas? They're very fierce. They'd eat you if they met you in the jungle.' I could not allow gorillas to be so slandered. 'They're quite peaceful really,' I said. 'They wouldn't eat you. They only eat leaves and things like that.' 'They would if you took their babies,' he said. 'They're very tall. Like this,' he said, standing on tiptoe and stretching. 'They'd push you over.' I walked absent-mindedly away, smiling, moving towards an interesting configuration of girls, and the sound of chanting, in the distance. '*Actually*,' said the voice at my elbow, 'gorillas are the strongest animals in the world.'

The three girls had stopped their chanting and told me the game was called 'Wash the dishes, Dry the dishes'. 'You have two people standing facing each other and holding hands, and another person in the middle. You say, "Wash the dishes," and you put your one lot of arms over her head. Then you say, "Dry the dishes," and you put your other lot of arms over her head in the other direction. Well, she is between your two lots of arms now—they are round her waist. Then when you say, "Turn the dishes over," you uncross your arms and you tumble the person right over on to their feet again, like a backwards somersault.' 'You can quite hurt yourself,' said one of the operators, rubbing her wrist.

A bullet-headed boy racketed round the playground asking all and sundry, 'Do you know a girl called Spaghetti?' and a girl hovered near him answering, with a kind of intoxication, '*I'm* not called Spaghetti.' It was a duo: a delight. This playground needs no visiting companies of clowns, we have our own.

The gorilla boy was still there. 'Elephants are *not* very strong,' he said. 'D'you know how a gorilla walks? *Stamp*;' and then, wistfully, 'I wish gorillas could talk.'

[1] Super-balls are the current craze. No bigger than a large marble, feather-light and tension-packed, they fly high in the sky looking like brightly coloured exotic birds.

In the 8-year-olds' queue the girl who had claimed 'Scribs' still had her fingers crossed 'so that they can't have me'. 'What have you been doing in school, so far? You came back on Tuesday, didn't you?' A boy answered for her. 'Same as we've always done—*work*,' he said with mock gloom. 'We've had tons of maths tests,' said the girl. 'We have to get 200 points.' 'And if you don't?' 'Oh I don't know. But if you do they say, "Well *done*!" '

After they had gone in I lingered with two girls who were finishing a clapping game:

> My father went to war, war, war,
> In nineteen seventy four, four, four,
> He brought me back a gun, gun, gun,
> And shot me in the tum, tum, tum.

Always something new in this playground.

Wednesday 13 *September*

The sky overcast, with glaring white light shining through. House-martins hunted the patch of sky directly overhead; then went on, leaving the sky empty.

I waited for the 7-year-olds to assemble in the old Infants' playground. As they crossed over the road in a frieze I noticed how their knees never quite straightened. Bent-legged childhood and bent-legged age, taking precautions against falling. An older girl mothered the newcomers as they arrived in the main playground. She took two of them by the hand and leant down to explain something. I overheard: 'Do you know what that lady does? She goes round asking what games you are playing.' I walked on, feeling less confident than usual.

A dreamy little boy, who thinks about things, was skating along with his right foot on a bottle cap. The crinkly edge of the bottle cap bit into the sole of his shoe, and the rounded surface slid on the ground like a furniture castor. It made a screeching noise. After a while he stopped, picked the bottle cap off his shoe, and handed it to me. 'You feel that,' he said, 'it's warm.' 'Do you do it because of the noise, or the sliding?' I asked. 'Both,' he said.

Two 8-year-old girls came and stood with me, smiling. They were eating apples. 'Everybody's eating apples this morning,' I said, looking round, 'there's even a core on the ground already.' 'Mm,' they said. 'Funny sort of morning,' I said, 'I don't know whether I'm hot or cold.' There was no response. The older girl came over and asked them, 'Do you want to play Stuck in the Mud?' They shook their heads. I was glad someone had mentioned games. 'What games have you been playing?' I asked. 'Yesterday for instance.' 'Yesterday—we played Chase on Lines.' 'And Red Rover,' said the second girl, 'and Sly Fox.' 'Three games. Not all in the morning?' 'No, in three parts of the day—morning, dinner time, and late afternoon.'

The gorilla boy was nattering quietly in the background. 'They're not very gentle, are they, gorillas? Where do they put their hands? Under their armpits, like this. Where do gorillas live? In zoos!' 'He's crazy about gorillas, that boy,' said one of the girls dispassionately.

A girl was being pulled two ways by two boys who each tugged at a sleeve of her jersey. 'Look out!' called out one of my companions, 'one of your arms is tearing!' Indeed one of the sleeves was ripping at the seam. 'What's happening?' I asked her. 'I dunno. We're playing Space. *He* wants me to go to prison, and *he* wants me to go on a space ship.' One of the contestants stopped pulling and asked me, 'What's white and goes up? A stupid snowflake.'

Rough and tumble seemed to be the order of the day. A boy jumped on to the back of the flat-faced girl. 'I don't know what's the matter,' she said. 'The boys have all gone mad. They keep trying to kiss me.' And she is such a solid, football-playing type, too.

However the boys were chiefly occupied with football (using smallish white plastic balls, with coloured patterns) and racing cars. Where the grass joins the tarmac a dust-bowl has slowly been growing. Once it was a series of shallow hollows used for marbles play. Now it is a depression deep enough for a platypus to nest in. A crowd of boys—at least ten—were hurling toy cars and lorries along the tarmac towards this cavity. The idea, I was assured, was 'to make it go along into the hole without toppling over'; but the enjoyment of the game seemed to derive from the speed and wildness of the throwing, and the quick retrieval of the vehicle after it had (as it usually had) landed on its back. As a boy positioned his car he shouted, 'Out the way, I'm having a go. Watch this then!' As he hurled the car he screamed, 'Neeow!' A boy who had no car was making do with a lozenge tin; it had a much more stable shape than a car and was in fact very much the shape of the Formula One racing cars, which are very wide. The casual, careless, endless crashing of the cars was disturbing; it too closely resembled the pile-up at the Italian Grand Prix this week, when Ronnie Peterson died.

A girl was squatting beside the car pit, eating 'bacon 'n bean' flavoured crisps. Sometimes a boy would call out, 'Hey, Nicola, did you see that?' I thought she must be filling the role of Queen of the Tournament, but she was watching the boys' technique intently. She turned a pair of serious eyes to me, reluctantly, and said, 'It's just that—well, I've got a load of cars at home, and I like playing with them. I've got a friend I play with—a boy. I don't bring the cars to school in case they get broken.'

In the reverse-busking of the waiting queues, when the queue does the entertaining, I learned some more riddles—even a 'true' riddle of the ancient and poetic kind, which is a rarity nowadays. The shyest, usually quite silent, 8-year-old had just learnt it from her 10-year-old brother. 'As light as a feather, As round as a ball, And 300 men cannot carry it.' The answer was of course 'A bubble'. The usual 'riddles' of the playground are conundrums; they have punning answers which are difficult to think of on the spur of the moment, and, unless the person asked has heard the riddle already, they usually constitute a comic double act:

'Why did Sir Wellington invent the Wellington boot?'
'I don't know. Why did Sir Wellington invent the Wellington boot?'
'Because he could smell de-feat.'

'Why is the Sahara a good place to have a picnic?'
'I don't know. Why is the Sahara a good place to have a picnic?'
'Because of the sand which is there.'

Repeating the question is an aid to learning the conundrum. It certainly helps me to get the wording right—the children correct me if I get it wrong.

Time for one more. A girl wanted to know why the chicken crossed the road. It was to get his pension, she said—'Do you get it?' Obediently, I said, 'No.' 'Neither did the chicken!' The ritual was complete.

Monday 25 *September*

The shocking-pink sunset last night (as well as the forecasters) had promised us brightness today; but at half-past ten the sky was still sulking greyly.

A boy was standing and writing in the middle of the playground. I felt a sense of kinship, and competition, and went and asked him what he was writing. 'Oh, car numbers,' he said, giving me an iron glance through his spectacles. 'I do cars from abroad.' I gathered that his was a solitary mania, and was glad when a curly-headed girl came up, sucking an orange, and said sociably, 'I'm freezing. I don't need a coat, but I really need a proper woolly.' We assessed the situation. She was wearing a summer vest under her white cotton shirt, and I suggested she button up her grey cardigan. The conversation was feminine and cosy.

'What games have there been lately?' I asked my humdrum companion. 'Nothing special,' she said in an offhand manner (and this would be my own summary of the games so far this term). We strolled over to where a game was brewing. Six or seven children were concentrating on each other, becoming active, becoming a self-reacting entity. Their faces were animated, they communicated with quick smiles. They started running in different directions. One of them shouted, 'Who's on it?' and another replied, 'Helen's on it.' 'I'm no-ot,' shouted Helen. The confusion about who was chaser made the game more fun: muddle is in itself intoxicating, and they laughed immoderately. A boy, meeting them head-on, was brought into the game. He ran away; then, realizing he had run beyond the boundaries of the game, ran back towards the others. 'Who's supposed to be on it now?' they called to each other, giggling. 'I think it's Nicky.' It seems that the basic games, though not 'special', are as good fun as ever.

The gorilla boy put his arm in mine and assured me that 'Gorillas can lift you up.' It is as if he is in love with gorillas; their name must always be on his lips.

The craze for miniature cars was still going on. Clusters of small boys were manoeuvring them up and down the hillocks at the edge of the grass patch and whizzing them across the tarmac. Two of them came up to me. 'Look what's happened,' said one, gleefully, holding up a tiny blue car with its roof squashed flat. ''E fell over, and then I fell over, and it got like this.' I said that anyway they had a wrecked car now, which they could not buy in a shop.

A vivacious group of nine girls (and a boy who had joined them by invitation) were playing Eggs and Bacon.[1] The leader began swinging the players round one by one and letting them go. 'Eggs, bacon, or round the world?' she asked the first girl in the line. 'Eggs' and 'Bacon' are slow, I was told, and 'Round the world ' is fast. Where the players landed, there they stood. Then the leader went round whispering to each. 'What did she whisper?' I asked one of the girls. 'Pop dancer,' she said, looking worried to death. 'And you've got to act it?' 'Yes.' The leader finished whispering and shouted 'Lights on!' All the actors acted. She shouted, 'Lights off!' and the actors crouched down. The leader went round and touched all except one; that one had to go into the middle of a circle of all the other players, who danced round chanting,

> Wake up, sleepy head,
> Don't forget to make the bed.

As soon as the rhyme was over, the one in the circle had to try to break out, and 'if they do, they're on it.' Later, when they had finished the game, I checked on some of the rules. 'Is it the best actor that she leaves—that she does not touch?' 'No, just

[1] See I. and P. Opie, *Children's Games in Street and Playground* (1969) 'Statues', 245–7.

anyone.' 'So it doesn't matter how well you act what you've been told to act?' 'No.' They thought I was mad, thinking it mattered; and I thought they had got the rules wrong and had destroyed the point of the game.

The bell must have rung because the football came to a halt and the footballers gathered by the classroom steps, still full of uproarious energy. 'Shall I tell you my nicest story?' said Paul, the lanky blond sophisticate. 'There was this lady and a man, and he said, "I'm going to get divorced from you, you don't half smell. You've got rubber lips and smelly breath. The rubber lips is your bottom and the smelly breath's your fart." I made that up.' He got his breath back and recited,

A man's occupation
Is to stick 'is cockeration
Up a lady's ventilation
To increase the population
Of the younger generation.

'I know one,' said Andrew.

There was a young man from Cosham,
Who took out his balls to wash 'em.
His wife said, 'Jack,
If you don't put 'em back
I will tread on the buggers and squash 'em.'

The polite version of that, which I have known since the 1930s, concerns eyeballs. I am not sure I don't prefer the rude one.

I was much too busy writing the rhymes down to register any expression. The boys were squealing slightly with excitement, and watching my face. 'They're good, these,' said Paul, with forthright approval. Another boy started,

Mary had a little lamb,
She thought it rather silly,
She—

The rest of the rhyme was lost in the flood of contributions, though I thought I heard the comfortingly juvenile term 'willy'. The next rhyme I could disentangle was,

There was an old man of Locket,
Who tried to invent a rocket.

> The rocket went bang,
> His balls went clang,
> And his dick ended up in his pocket.

'You'd better go in,' I said, 'the bell went a long time ago.' They scampered in, still squealing a little. 'We are a virile nation,' I murmured. I wondered whether football really sublimates the sexual urge, as the Victorians believed; and I remembered that the songs sung in pubs after rugger matches are scarcely reticent—but perhaps that is part of the sublimation process.

As I came up to the 7-year-olds' queue by the gate I heard two of them clapping 'A sailor went to sea, sea, sea' with crisp expertise; and saw that several boys and one girl were manipulating paper 'salt-cellar' fortune tellers. 'What do you call those?' I enquired. 'They're called Crabs.' 'Spiders,' said one boy on the fringe. 'They're *Crabs*. He's stupid. He's stupid in class. Go away!'

Monday 2 *October*

The weather has repented after the bad temper of the past few days. This morning was all soft high clouds and sunshine—almost hot.

The children came out at their leisure, eating apples. Some of the girls set up a skipping game across the front of the school, with a long, nearly uncontrollable rope. The senior girls began playing high-backed leap-frog; and the boys were tossing new yellow tennis balls. Perhaps they are not allowed to play football at the moment because the painters are painting the protective frames, covered with wire-netting, that fit into the school windows.

Two girls in the corner seemed to be receiving instructions from a third (a blonde, with a full, earnest face). One of them was folding her arms and saying, 'Yes, I've learnt this game. I *know* how to do it.' The blonde began slapping their outstretched palms and saying,

> Anna banna boo and a wheezy anna,
> My black cat can play the pianner;

He can play for two and a tanner;
Which would you prefer—Fish, Fight, or Clock?

They milled around for a while. I could not understand what was happening, so I had to ask them to explain afterwards. The blonde bent her head over my notebook. 'Have you got the words? ". . . play the pianner." Then, I don't know what it means, but it goes, "He can play for two and a tanner." ' I told her that a tanner was a sixpence, and she was interested. One of the others chipped in, 'The game is called My Black Cat.' The blonde continued, 'They hold out their hands and you slap them and say the rhyme and whoever's hand it finishes on you ask them, "Which would you prefer—Fish, Fight, or Clock?" and they choose. "Fish" is, you go like this [she raised her arms alternately] and I hold my arms out and you have to get under them without touching them. "Fight" is, you have to hop round with your arms folded and fight each other. "Clock" is, I have to draw a number on her back and she has to guess it.' 'I've never seen that game in this playground before,' I said. 'It's like a racing game called Kerb or Wall.[1] It has the same rhyme.' 'It comes from the north,' she said, 'from my other school, in Manchester. I've been at this school a week and two days.' As I moved away I heard the game continuing. The blonde was drawing on a girl's back, and when she could not hear her guess she came round in front of her and said, 'Pardin?' (All the children say, 'Pardin?', these days, whether they come from the north or not.)

I had seen, with pleasure and relief, a boy standing in the brave stance of a conker player, his conker held out on its string for another player to hit. I was content to watch, not interrupting. The conkers are particularly fine this year. His was strung on a new black shoe-lace. I was almost annoyed when some friendly lads came over and one of them asked me, 'What do you call two rows of cabbages? A dual cabbage way.' They had stopped in the middle of a game to tell me, too. The riddler glanced towards the far side of the playground and said with amusement, 'Tim's still running. He doesn't know we've

[1] See I. and P. Opie, *Children's Games in Street and Playground*, 196–7.

stopped. We're playing British Bulldog.' They went back to their game.

Conkers is a spectator sport. The heroes in combat were surrounded by an admiring double circle. I told the boy next to me, 'We were given a 2,107-er the other day, by an old man, from his schooldays in 1909. It was tiny, like that. All shrunk up.' He looked at me with glowing eyes, and drew in his breath. 'Oh, how marvellous! You'd never win against that, it's so small.' We were silent, imagining the age and glory of the ancient conker. 'I've got a small one,' he said. 'I broke one of me brother's 62-ers with it. I left it in vinegar, and it shrank. You should really only leave them in for half an hour and then dry them for twenty-four hours.'

Several boys were agitating to tell stories, eager for their chance of immortality. 'Let's have your story,' I said to a solemn boy with a long yellow forelock. He looked important and announced 'The title is called, "The Three Windows".' I do not recall stories with titles before. 'There was this man who 'ad to clean the windows on a house, and the boss said, "Don't clean the last three windows." So 'e cleaned the windows and when 'e came to the last three windows 'e said to 'imself, "They look so dirty I'll clean 'em." 'E looked through the first window and saw these willies on a table wiv a mouse in the middle. And the next window, there was a man hanging out by his dick. And the third window 'e saw this man taking off 'is rubber Johnny. And 'e asked the man, "What was it, in the first window?" and the man said he had picked the mouse off of a lady. And 'e asked, "What was in the second window?" and 'e said 'e'd put some cheese under 'is dick.' By this time the noise was so great and the story so confused I had to ask for quiet and some explanations. 'Why was he hanging out of the window?' 'It was a punishment.' 'And the third window?' ''E'd just 'ad it off. You know, with the lady. Then that's the end. You got to write down "Finished".'

'Whoo, whoo, have you ever seen a monkey do that?' A small figure in a green jersey was beating its chest by my side. 'I haven't seen a monkey do it,' he replied to his own question, 'but I've seen a gorilla do it.' Luckily the bell had gone and it was time to join one of the queues.

I asked the conker players how far they had got, and who was the conqueror. 'We haven't got very far yet,' they said. 'We only started playing about Friday. His is an 11-er—that's the best.'

Wednesday 11 *October*

A soft hazy day—warm for the time of year. There was a heavy dew early this morning. I walked to the school slowly, taking the short cut behind the inn, watching the young mothers wheeling their babies in baby-buggies through the pools of Spanish wine spilt by the delivery man.

Two boys appeared in the main entrance, holding the books they had been using in class. They stopped just outside, toed an imaginary line, and made a race out of the distance back to their own classroom. In the 7-year-olds' playground the children were racing to get a good place in the line which would be marched across the road. We are born with a conviction we must be first, which Christ and his epigrams can never overcome.

The 7-year-olds came over, clutching apples and Smurfs of different sizes (Smurfs are fantastical TV cartoon characters, now marketed strongly); also conkers on short lengths of string. A girl began skipping with a rocking-horse motion in a single rope, but casually, not as if skipping was an important game. I went to inspect the builders' site-hut, which I had not noticed before. It has become part of the geography of the playground. Two of the smaller boys were crouching behind it, eating their crisps in peace.

Between the school field and the back gardens of the neighbouring houses there is a fence of wire and wooden stakes. This morning there were cobwebs, spangled with drops of mist, hanging in every space between the stakes. Little boys were walking beside the fence collecting the cobwebs on loops made of tough grass stems (one of them had bent the stalk of a nettle into a loop). They lifted the webs with one upward stroke, as if they were ladling water; the idea was to get layer upon layer of cobwebs on the same loop of grass. As he walked back to the school building a boy with only one web on his loop was carefully spreading the drops of water with his finger so that a thin film of water covered the whole web: 'Look, it's like a mirror!' he said in an awed voice. (The headmaster told me he did the same, as a child, but he scooped up the cobwebs on a loop of hazel.)

The playground was mellow and relaxed, in this time of grace before winter. It seemed a time for minor amusements, rather than vigorous games (though football continues unabated). A girl with her eyes closed was being guided along by two others. They stopped when they met someone. Her hands were placed on that person and, keeping her eyes closed, she had to say whether she loved, liked, or hated them. Then she had to feel them all over and guess who they were.

Senior girls, propped against the south wall eating crisps and ignoring the footballs slamming the wall around them, said wearily, 'There's nothing much going on. Nothing, really.' And yet the west side was thick with games. I tried to reach the row of skipping games, and found I was getting in the way of girls with determined faces who were playing Sly Fox. A charitable 'ender' tossed the names of the skipping games over her shoulder: 'That one's Higher and Higher, and our one's High, Low, Dolly, Pepper, and over there's Building Up Bricks.' All the games were well attended, though over at the front of the school I could see a huge circle of girls, with an outer ring of spectators, jinking to the movements of Shirley Temple.

I reached the circle just as the bell rang. 'There must be at least twenty people playing!' I gasped to the girls, who had stopped playing but did not feel it was necessary to go and line up yet. 'Yes,' they said enthusiastically, 'it started up at the Infants. They were doing something from Shirley Temple and they made up things about her, what she looked like and things like that. See, she used to wear short dresses, so they made up that bit about her.' 'We are playing it next playtime, too,' said one *aficionada*, thrusting her head into the group. The game was marvellous; the story of its arrival was marvellous. They would not have thanked me for telling them that the game had a similarly miraculous advent in the winter of 1969, when it was brought by a new girl from Wales. The words now are:

I'm Shirley Temple,
The girl with curly hair,
I've got two dimples
And wear my skirts up there.
But I'm your nable
To do the sexy cable,
I'm Shirley Temple
The girl with curly hair.
Salami, Salami,
You should see Salami,
Hands up here, skirts up there,
You should see Salami.[1]

'But what does "I'm your nable" mean?' The girl who was dictating the words screwed her mouth up and raised her shoulders so high they almost touched her ears.

Walking back to the gate I passed the gorilla boy, who beat his chest silently at me as I went by.

The 7-year-olds, already under the control of the teachers ('Come on, two lines! *Mandy*, you leave John alone!') were bubbling with excitement over a large white van, decorated like a children's picture book with penguins in several bright colours, which was parked outside the school. 'Do you know what is in that?' one of them asked me. '*Crisps.*' The boy at the head of the linc asked his teacher, 'If that van says "Penguins" on the outside, how come it's got Outer Spacer crisps inside?' 'I don't know,' said the teacher, and they all went over the road.

Wednesday 18 *October*

Bright brisk October weather, shiny as a conker.

The 7-year-olds come across the road in classfuls. In one class, already lined up and waiting for its teacher, two girls were disentangling and rewinding a length of blue plastic clothes-line. I wondered if it was too thin for skipping. Behind me a long-drawn-out 'Ee-ow!' signalled the emergence of an 8-year-old into the main playground. I turned and watched him run to

[1] See I. and P. Opie, *The Singing Game*, 417–19.

the gate and swing on it. The gate is the right size and shape for swinging on: eight iron stakes and two iron cross-pieces, the bottom cross-piece being right for the feet, the tops of the stakes right for the hands. 'Is it a good gate?' I asked him, listening to the gate creaking and wondering what answer would result from such a vague question. He squared up bravely and said, playing the part of Jack the Giant Killer, 'Yes. But the thing is, I can't do a karate on it or it falls off,' and chopped the gate with the side of his hand. All the courage and resourcefulness and absurdity of the pint-sized hero were in that reply.

Behind him, other heroes were practising their conker strokes. County cricketers or golfing champions could not have discussed more seriously the arc-like movement of arms through air. 'Right,' said one at last, 'now you hit mine.'

A chatty little girl tagged on and walked round with me. 'Any new games?' I said. 'All the same games,' she replied, 'except conkers.' 'Yes, the boys are playing conkers now, aren't they,' I said, glancing back at the conker players. When I looked at her again I saw that, miraculously, her hands were full of conkers, and she was pulling more and more from the recesses of her skirt. 'You know how big the school desks are?' she said. 'Well, mine is full of conkers. You can play marbles for conkers. Oh yes, I play conkers against the boys—knock 'em out.' Nevertheless she was occupied in talking to me, and there was not a girl conker-player in sight.

I saw that the blue plastic clothes-line was being used for skipping. Seven girls were playing 'Apple, apple' (the perennial stand-by here). Further down the east side another seven girls were playing 'Down the Mississippi' in a very long, real rope. Jack the Giant Killer had attached himself to me and said, with transfixing blue eyes wide open, 'Do you know my favourite hobby? Watching TV! He he he he!!' (He must have known how irritating this pronouncement would be to a devout folklorist.) 'Yes,' he said. 'My best programmes are Kojak, Starsky and Hutch, Columbo, James Bond' (seeing that I was having difficulty in keeping up he said, 'Just write down "007" for Bond, and then make the 7 into a gun, because he's a detective—they're all detectives'). The detective is the present day super-hero.

I tried to watch the skippers, pleased that skipping seems to have arrived at last. The Giant Killer seized the end of the plastic skipping line. 'Leave go, Chris-to-*pher*!' yelled the girls. He stood in front of me again and said, 'Do you want to know what else I do with my life? I usually run round the Close five times a day—or ten times.' '*And* he's got a train set,' said an admiring friend.

A girl was walking across the playground lifting her right foot along with the aid of her single skipping rope. Suddenly I remembered that I used to do that when I was a child. It is a marvellous mechanism. You must hold the rope taut, at waist level, and put your foot on it, and when you walk the foot is carried along at every other step without any effort on your part.

A crowd of boys was gazing at the school roof. A tennis ball or a conker was stuck up there, no one knew which. The Giant Killer was still trying to impress me. 'Do you know what I did yesterday? I made a mud pie and put it on a plate and pushed it in my Dad's face. He had to go and wash. And I smash windows—with gloves on, of course.' His final burst of creative energy came in the form of a riddle: 'What's hairy and has 569 legs and can't move an inch?' 'I don't know. What is the answer?' 'I don't know, either.'

The crowd had gradually drifted away, but the half-dozen boys who remained felt that some entertainment was called for. 'Do you know the one about the Irishmen they wanted to kill, and they gave them one last request?' 'That's too long,' someone said, but the teller had already started. 'The first one said he wanted a bottle of Scotch, and the second one said he wanted a bottle of whiskey, and the last one said he wanted a piano. So they put them in the gas chambers, and when they went back two hours later they found the first Irishman in the first chamber and he was dead. And they found the second Irishman in the second chamber and he was dead. But the third Irishman in the third gas chamber was alive. They said to him, "How did you survive?" and he said, "Tunes make you breathe more easily."' ('Tunes' are cough, or asthma, sweets, advertised with this slogan on television.)

We were lost to the world by this time. 'Have you heard about the Irish Humpty Dumpty? The wall fell on *him*.' 'Then what about this one?' shouted another, dancing all round me,

> Humpty Dumpty sat on a wall,
> Eating black bananas,
> Where do you think he put the skins?
> Down the King's pyjamas.

I was surrounded by leaping boys shouting, 'I know one! I know one!' Then, rather to my relief, the bell rang.

A 9-year-old girl in the line-up claimed to be as tall as me. She was mounted on a friend's back, and both of them wore gold ear-rings. Sometimes I have the feeling I am in Italy. 'Everybody's wearing ear-rings,' I said dolefully. 'Yes, but mine are horse-shoes,' said the rider, clinching some unspoken argument. Over in the parallel queue two girls were kneeling, each with one knee raised, and hand-clapping. They looked like slave girls in an ancient Egyptian frieze. They were clapping to an exotic chant:

> Em pom pee diddy vee diddy voskus,
> Em pom pee, diddy vee;
> Diddy voskus, diddy voskus, diddy voskus—
> Poof poof!

Then they saw they were late and, with a last great shove, fled into school.

Thursday 26 October

Low grey scudding clouds. Sparrows flying their jerky, hurried flight, as if trying to escape out of the sleeve-end of the world before the clouds caught them.

A girl with the round, wholesome, curl-framed face of a 1930s bride-to-be in *Country Life* stood waiting by the gate. We engaged in polite conversation. 'Quite a lot of skipping now, isn't there?' I said. 'Yes. I'm waiting for a friend of mine—she's a first-year. She's got a rope, a very long one.' The conversation flagged. 'Her name is Helen,' said my companion, and went to hang on the fence, looking across the road into the 7-year-olds' playground. 'How brilliant, she's got her rope!' she called out, then, turning to another girl, asked her, 'Are you waiting for who I'm waiting for?'

Jack the Giant Killer arrived, walking in one of the limitless variety of ways possible to children. He was stuttering along with his feet, and a friend was copying him. 'Stop it!' said the Giant Killer, and clamped the friend to a standstill. 'Would you like a joke?' he said to me. 'Why couldn't the skeleton go to the dance? Because he had no body to go with.' 'No body—no body,' repeated the children standing near, in an appreciative murmur. 'I've got a skeleton story, but I'm not very good at telling stories,' said a small girl, eyes cast down under a dark fringe. She told it very quickly. 'There were these two skeletons and one was older than the other. The older skeleton said, "I dare you to jump in the river," and the young one said, "I haven't got any guts."' Then she ran to join the skipping game.

There were twelve people in the game already. They were playing an inferior form of 'All in together girls, When it is your birthday please jump in'. The 'very long rope', much coveted and long awaited, turned out to be five pieces of disparate, ragged, thickish string knotted together, which shows that home-made playthings can still be prized in an age of affluence. The rope, despite its thinness and extreme length, was functioning extraordinarily well. It was the skipping that was at fault. Before all the months had been called someone invariably tripped, and the game had to be started again. I wondered if they would ever reach the stage when, the rope filled with all the players, the sequel could begin and the players be told '. . . When it is your birthday please jump *out*'. But soon, anyway, the boys began attacking the rope and the whole game had to be moved to the front of the school.

The Giant Killer caught me up. 'Do you know there are some new posters out, with a picture of you on them?' he said boldly. My mind raced through several possibilities. Not even he could have mixed me up with the Conservative poster of Mrs Thatcher at the crossing-gates. 'No. You're joking,' I said. 'It must be one with Mickey Mouse on it.' He grinned and pointed at a sticker on a friend's jersey. Round the three-inch wide sticker was printed 'The Football Crazy Gang, Smith's Crisps', and in the centre a particular member of the gang was celebrated—Spotty, in this case. They showed me another (Chopper, equally repulsive), and gave me a potato 'football' from the packet.

On the west side of the school the ropes were turning, in rows, as if in some long workshop full of machinery. Orders were issued: 'Vicky, get behind me—go on!' 'What are we playing, then?' The birthday game seemed the most popular. And where there is activity there are spectators. An equal number of girls were standing, as I was, just looking at the skipping. Alongside the row of skipping ropes, boys were playing Stinger against the wall, with no discernible boundary between one game and the other.

Idling, at the end of playtime, I asked a girl what other games were going on. 'Conkers,' she said, blowing up her potato-crisp bag and bursting it. 'I saw a boy yesterday', she said, 'with twenty-six conkers on a string, and he was playing with the one on the end of the string, and if it was a one-er and he lost it he just moved another one down the string.' 'My brother plays conkers all the time,' said a tousled imp. 'He's a little *baby*, he's fifteen.'

Thursday 9 *November*

A chill St Martin's mist. Everyone in the village, it seems, is wearing a poppy.

I felt quite strange in the playground after a fortnight's absence (last week was half-term). I noticed, all the more, the busy environment the children create for themselves, flying in all directions like bees in a bottle, intent on a hundred and one trivialities: walking quickly heel-and-toe, pointing two-finger guns, peering round corners, hurling toy cars, riding piggy-back, juggling bags of crisps in the air. Innumerable tussles and brief competitions go on all the time. Two girls were holding each other's shoulders and trying to tread on each other's toes. Another pair, looking rather self-conscious and, I think, watching to see if I was looking their way, kept on measuring up to each other and lifting and dropping their arms. 'What is it you're doing?' I asked, as I was expected to. 'It's a piece of judo,' they said.

A little boy was kneeling on the ground, setting off a strip of caps, one by one, with a stone. A second boy was examining the Clark's [Shoes] Commando Chart for which he had swapped the strip of caps, and at the same time keeping an eye on his former possession: 'Don't use them all up,' he said, 'or you'll have none left.' The Commando Chart gave tracking signs (including a Clark's shoe that could be placed in a significant manner) and hand signs for inter-Commando communication.

Boys in the distance, who were standing and eating apples, suddenly changed into a posse of armed men and charged round the corner with guns firing. There was a lot of shooting in the playground this morning. Lonely little boys, apparently rejected by their schoolfellows, turned out to be part of the same game. I asked one who was standing stiffly by the wall, 'Are you playing anything?' (knowing it was a silly question as I said it). He said, 'No. Yes, I'm 'iding from someone.'

I found I was standing next to the mouse boy (so called because he reminds me of the sharp-faced, intelligent mouse in the *Daily Mail* strip cartoon I used to read as a child). 'You wouldn't think they'd want to play those shooting games, would you?' he said. 'They're just a mess. I'd rather do karate myself.'

On the east side, girls were scuffling for position in a queue, waiting their turn in a game I had not seen before. It involved throwing a large yellow dice into the circle of the netball court. 'I'll explain it to you,' said a heavily built girl (shall I call her 'the tank girl'?) gallantly relinquishing her place in the queue. 'You see, there's somebody standing there by the circle, and there's somebody at the top of the queue. Well, they chuck the dice and then whoever gets the highest number has to chase the other one up to that post and back, and if they catch them they go to the back of the queue.' 'And if they don't catch them?' 'They change places. The one who was out in front goes to the back of the queue.' It seemed a good game. 'If you both get a six you don't run,' she said. 'And if you both get a one, do you run?' 'No.' (Much probing is needed, to discover a general rule.) 'You have to throw again,' she added helpfully. 'It's a huge dice,' I observed, surprised to find that I felt slightly shocked to see a dice in the playground. 'It's a rubber,' she said. 'You get them at Menzies for 10p. Oh, six *again*, Cheryl!'

'There's another game we play with dice,' they said, crowding round. 'We play it in the toilets. Someone goes into a toilet and locks the door and there's a whole queue of people outside, and the person inside, they throw the dice out through the top of the door where there's a gap and—there are these kind of squares outside—say it hits the third square along, you count [i.e. score] three. But if it hits somebody on the head you try again.' Apparently they were not concerned that the dice was not being used as a dice, and that any object could have been used for throwing out on to the squares. Possibly the game started with throwing the dice out and scoring whatever number was uppermost, and then evolved into the present game.

'What about that game with a rubber octopus?' shouted the tank girl. 'You put the octopus on a drain. We'll show you. You get in a circle—Marlene, get in a circle!—and you kind of pull each other on to the drain, and as soon as you tread on the drain you have to go and be tied up on the railings—well, you're not actually tied up, you just put your hands behind your back. And as soon as somebody else is out you can come back.' They were circling and shoving and laughing, like Peter Pan, with the joy of their own cleverness. 'They're all new games,' they crowed. They were so nearly right that I refrained from asking them if they had heard of a game called Poison Pot, or This Drain is Poisonous which was known in various places in the 1950s and has been known under various names since at least the 1890s.[1]

Some of the boys had been playing Om Pom, yelling and hiding behind the dustbins. The two naughty boys (thank goodness there are never more than two or three like this) detached themselves from the game in order to bait me. They followed me around, sometimes almost serious, sometimes unhinged by hilarity. 'Hey, are you married?' 'Hey, are you married to a millionaire?' (Each elaborates the other's remarks.) The finale was, 'I know a game—Dog's Muck.' 'Yeh, what

[1] See I. and P. Opie, *Children's Games in Street and Playground*, 237.

comes out of yer bum.' ''Ave you written that down? You *never*!' They are, I've noticed before, almost unable to control what they are saying, and yet they are half-interested that I am writing it down, too. One on his own would never be so daring, indeed the aim is to be daring in front of a witness—to show off, in fact. 'Don't you know any real games to tell me?' I said. They shaped themselves into a Camel, and gaily went off across the playground.

'Has the skipping stopped?' I asked the girls, remembering the thriving skipping craze of the week before last. 'Yeh. Too cold for skipping,' said the tank girl confidently. 'Your hands would go tremble tremble when you are turning the rope.' 'It's time for netball now,' said another girl. 'We've got a lot of matches and we practise every day.' I suspect that the half-term holiday killed the skipping craze, though I noticed a few single ropes being carried into school, as I went away.

Tuesday 14 *November*

Dark rain-laden clouds racing across the sky. 'Blowing all over the place,' they said in the village. 'Doesn't know *what* it wants to do.'

I was 'bagged' as soon as I set foot in the playground. A sharp-featured, bright child of about nine, with a blonde fringe and gold ear-rings, came racing to book my services as amanuensis. 'Do you want to write down my joke?' she gasped. 'Adam and Eve went out in the garden. Then Eve went in to make the dinner. When the dinner was ready Adam came in and said, "Eve, you've put my best suit in the salad again."' It was a sparkling performance, and I felt she was justified in looking round at her friends for admiration. 'It comes off Basil Brush, on the TV,' she said, 'Every Saturday, about five I think, quarter to or quarter past.' She ran off to join her coterie, having raised her prestige enormously. It is odd how many girls in this playground have sharp features, bright eyes, blonde fringes, and gold ear-rings, especially those with histrionic abilities.

A dustbin had been overturned, or a refuse cart had overflowed, near the school fence; there was a band of litter several yards long—scraps of paper, sweet wrappers, and crumpled newspaper. The naughty boys were gazing at it. 'You know what it is?' said one of them, turning to me. 'It's what we said last week, *you* know, from the girls' fannies. I expect you know what a fanny is.' 'Do you know what a *vagina* is?' said the other. 'It's part of a woman,' I replied. 'Do you know what a penis is?' asked the first. 'It's part of a man.' (The answers were accepted: I felt I had passed an exam.) They told me they were not planning to play anything. 'We're just goin' to muck around—chase the girls.'

The mouse boy and some friends were posturing like the wild dancers in Van Meckenhem's fifteenth-century *Ring Dance*, bending backwards to hold a kicking leg in the air. A boy accompanied them, consulting an open book. They were practising karate kicks, from a book entitled *Karate Test Teaching*, and were too busy to be interrupted.

Two little girls were passing the time by clapping 'Popeye the sailors' man'. Even while they were clapping, others were gathering round them waiting to absorb them into a game of Stuck in the Mud; and soon all that corner of the playground was alive with screaming, pouncing players, yelling, 'You're stuck in the mud,' and 'You can't do that.' Stuck in the Mud, like most good games, is open to minor deviations from the rules. It is an occasion for furious, rollicking disputation, the very spice of juvenile life. One girl, half-way under a prisoner's legs, was waiting for the chaser to become bored and go away, so that she could complete the rescue. Another girl had jumped on to a stuck-girl's back, instead of going through her legs in the orthodox fashion, and the stuck one, blissfully happy, was shouting, '*No*! Get *off*!'

Some of the younger girls have gone on bringing their single skipping ropes to school, although the main craze is over. A skipping rope is a companion. A child need not feel self-conscious if no one invites her into a game; she can be seen to have an occupation, and can offer an occupation to a friend. This morning two girls were skipping neatly in a single rope: the little routine consisted of turning into the rope one after the other to skip once, then twice, then three times. Another double act nearby, which the girls said they 'sort of made up', was 'Two little dicky birds', in which the birds started side by side in the rope and stepped nimbly out as they 'flew away'.

The tank girl ran up holding a fivepenny sausage-balloon. She blew it up and held it high, waiting for an audience to gather before she let it go. 'It makes a farting noise,' whispered a spectator to me.

The naughty boys crowded round like a football scrum. 'What's that about farting?' said one. 'You know how cowboys scoff beans? Well, after, they're cocking their legs and doing farting.' Merriment swept them into a whirlwind. They swirled round, blowing raspberries. A curly-headed youth began telling a story. 'There was this man kept on eating beans and farting, and he wanted to get married. So he asked this woman and she said, "I'll marry you if you promise to give up eating beans and farting." So he said, "All right." So on their twenty-fifth wedding anniversary he went to buy her some chocolates and flowers, and as he walked by the supermarket he saw they'd got beans half-price. He thought, "I haven't had a can of beans for twenty-five years. I think I'll have one today." So he went in and bought one and ate it down, and he didn't fart, so he thought, "I'm cured. I'll buy a hundred cans." So he did, and 'e ate 'em all down and 'e walked home. And as 'e walked up the garden path 'e felt this rumbling inside 'im and 'e pinched 'is bottom together and rang the bell. And his wife opened the door and said, "I've got a surprise for you."' ('I know what,' said one of the listeners, enthralled.) 'So she put a blindfold on him and led him into the front room. Then the phone rang and she went to answer it, and he did another fart, and he was waving his arms about to try and get rid of the smell, and she came back in the sitting room and took off the blindfold and there was all the family sitting round the room.'[1] I was exhausted. 'That's a long story,' I said admiringly. 'Yes, isn't it,' he said, looking proud.

Their teacher arrived and overheard me saying to the boys, 'Oh, you'd better go in.' 'That's all right,' she called cheerily,

[1] This 'modern legend' has been around for a long time. Carson McCullers included the story of the blindfolded farter in a room full of people gathered to celebrate his birthday in *The Heart is a Lonely Hunter* (1940). See Jan Harold Brunvand, *The Vanishing Hitch-hiker* (1981), 148.

'just keep the relevant people—the ones who are telling you something.' They all wanted to be relevant. 'One more joke,' I said firmly. The dominant story-teller decided on a short one: 'Who are the people who have the most friends to dinner? Cannibals.'

Tuesday 21 *November*

A mild grey day, yet a wind strong enough to strip the trees of their remaining leaves.

I leant against the telegraph pole outside the school, taking a long-distance view of the children in the playground. The skipping season is apparently still with us, after all, though the skippers are mostly separate and serious rather than communal and joyful. A tall, athletic figure in red jersey and trousers was skipping hard in a single rope; and others, less athletic and impeded by drooping skirts, were trying to master the trick of turning inwards into a single rope two at a time. Four girls were playing 'Down the Mississippi where the boats go push' in an inadequate length of clothes-line. Further away, pairs of scissoring feet told me that Cat's Got the Measles is back in fashion.

The game of Cat's Got the Measles seemed to be unusually lively, and as I drew near I understood why. Five pairs of white socks and two pairs of red socks were criss-crossing enthusiastically to a version of the rhyme which is new to this playground:

> The cat's got the measles,
> The dog's got the flu,
> The chicken-pox, the chicken-pox,
> And so have you.

For the past forty years the children in this playground have sung, to the tune of the 'Keel Row':

> The cat's got the measles,
> The measles, the measles,
> The cat's got the measles,
> The measles got the cat.

The introduction of another animal and two extra diseases has had the vivifying effect of a new spring fashion. The game soon grew to twice the size and had to be moved to a larger site. The organizers gave the players leave to run to the cloakroom for coats and scarves, so there would be more clothes to take off as forfeits. Then they began again, lumping their ungainly 7-year-old bodies (the despair, no doubt, of ballet-minded mothers) to words which have survived in one form or another through nearly a century of childish misunderstandings. They scorned to work out how to jump so that their legs would be crossed on the final word, thus avoiding the penalty of unzipping a jacket or throwing a garment into the ring. In fact the penalty seemed to be enjoyed. As I left I heard the boss of the game saying unctuously, 'I've got to take something down a bit', as she pushed her knee-high socks down to the ankle.

The marbles fence was deserted, as it has been all this term. The fence served only to support one end of a skipping rope for a modest game of Higher and Higher. The turner holding the other end showed me her two grazed knees (fustian skin, blood-speckled). 'The person who was holding the rope didn't let go,' she said. 'They never do. So I fell and I didn't feel like jumping any more.'

The fringed and ear-ringed girl (whose laughing eyes and chorus girl features I am beginning to recognize) came with a friend to tell me that they had been playing The Bird Game. She offered me a crisp and stood back while the more articulate friend (pale early-Victorian face almost hidden in the hood of her duffle coat) described the game which was 'Not much of a game really. You just jump off that sloping ledge over there and flap your wings. You have to do different movements, though, before you flap your wings. There's "the Loving Heart"—you put your hands together in front, and make a heart shape in the air. Then there's "the Bicycle", when you have to pedal. And "the Conductor"—you have to conduct. And "the Propeller"—you have to do the movement *and* fly before you reach the ground.' The chorus girl said disconsolately, 'I wish we could really fly though. I'm sure we ought to be able to. We would be able to if we had feathers.'

'What else has been happening? You don't *have* to play games in playtime, do you?' 'No, you can just walk around,' said the duffle-coated girl. 'And,' she said, looking at the chorus girl, '*she's* been having to write out spelling words, because she's been making people's names funny with the alphabet, and giggling in class.' 'Funny?' 'Yes. Well, supposing someone's name is Jane Lewis, she says, "Tane Dewis", and Rebecca comes out "Tedecca".'

I was not to escape the naughty gang, who came reeling round the corner when the bell rang. 'Curly wurlies!' they shouted. 'Dizzy wizzies!' With this lot, one is not entirely sure whether they are volunteering the name of a game or something more dubious. Could they mean 'pubic hairs'? I decided not to commit myself. They repeated 'Curly wurlies!' and went whirling off in a circle with their arms stretched sideways. 'Do you ever make yourself so dizzy you fall down?' I asked. 'Oh yes, *I* did. I fell down on my arse-hole. You written that down? You haven't got very tidy writing, have you? Write down British Bulldog, that's what we bin playin'. Write down Stinger, that's when you run across and they try to hit you with a ball. It's painful.' I was rescued by a civilized child who told me about a poem she had sent in for a Pony Club competition. It was a limerick, neatly rhymed, about a pony called Bill who went up a hill.

There was the usual rush to put the latest rhymes, songs, and jokes on record before going in to school. 'I gotta new rhyme! It's a sort of song, really!'

> One banana, two banana, three banana, four,
> Fifteen skinheads knocking at the door,
> Five with machine guns, five with sticks,
> Five with hand-grenades hanging from their—
> la la, la la la la la . . .

They all chorused the wedding march, and luckily there was no interval before the next song. Two girls sang, to the tune of 'Frère Jacques',

> School dinners, school dinners,
> Concrete chips, concrete chips,
> Soggy semolina, soggy semolina,
> Doctor, quick! I feel sick.

'We made it up in the back of the coach,' they said, 'when we were going to see *Noye's Fludde* in Basingstoke.' Then they gave a guilty look at the classroom window. 'Ooh, we ought to go in,' they said, 'we're doing tests.'

Tuesday 28 *November*

A malicious icy wind, in cruel contrast to the tender blue sky. I went into the school and looked at the library books in the passage until playtime began, and was grateful for the County Council heating.

But the children came into the playground cheering, and instantly started sliding on the ice rink that had providentially appeared in the shadow of the school. A boy came up to me, his happiness overflowing into words: 'Lovely skidding weather, isn't it? I like bombing down the runs. It's *good*, though, isn't it? I'm going to go on doing it now.' No one could have been more radiantly thankful for a gift from God. Another boy stepped forward. 'I *always* like skidding,' he said, with pleasure that must have been remembered from last winter, and rejoined the milling throng. I wondered if everyone called it skidding and not sliding. 'I don't know,' they answered severally, 'everyone seems to;' and they went on skimming past, content with provisional pleasures.

Soon, however, one of the girls was in tears. She had fallen while sliding, and had scraped raw a large area of bare skin on her thigh. No wonder only a few girls go sliding, when they are handicapped from the start by short skirts and bare legs. Most of the girls were taking the safer course of squatting and being pulled along by two friends.

A girl with a single skipping rope ran by, saying, 'I'm not staying there in case I slip,' and went round into the sunshine. Again this week there were only short ropes to be seen. I heard 'Two little dicky birds' being intoned, twice, and stopped beside the second pair of girls. 'Do you remember when that was just a rhyme?' I asked the small laughing one. 'Oh yes, and I know where the tiddly little dicky birds go to, as well. They don't fly away really, you just keep them under your fingers.

My sister told me the secret.' She and her friend had lost interest in skipping. They wanted to chat.

'I've got some lovely jokes. What did the bell say when it came out of the river? I'm ringing wet! and What dog can jump higher than a house? Any dog. A house can't jump! and How Hi is a Chinaman?' 'I know that one,' I said unkindly. I was not going to waste my playtime on 'How Hi is a Chinaman?' The little girl was not at all put out. 'Oh, I know lots,' she said ecstatically, 'I'm full of them,' and she went on pell-mell: 'Why was Cinderella thrown into the hockey team? Because she ran away from the ball.' 'But,' I said, 'shouldn't it be "thrown *out* of the team"?' 'No, *into* the team.' That was how it had to be. The other girl said, 'Perhaps she was thrown into the team because she was a coward.' We had to leave it, the jokes were still in flood.

The other girl wanted her turn. 'What goes along the bottom of the river at fifty miles an hour? A motor-pike and side-carp.' 'What's green and hairy and goes up and down? A gooseberry in a lift' (This is probably the Ur 'lift joke'). 'We mustn't use up all your paper,' she said proudly, as I flipped over the pages of my notebook. 'Just a minute, I'm going to tell one to Samantha and then she can tell you.' She took Samantha aside and whispered in her ear. Samantha came forward and said carefully, 'What is the difference between a jeweller and a gaoler?' and, having waited for me to repeat the question in the traditional way (which is always helpful to the questioner), said with satisfaction, 'The answer *is*, "One sells watches and the other watches cells."'

A very strong paper dart struck me in the temple: made from a paper towel, the boys said. 'But we are really playing Detectives,' they said solemnly, looking at me down the tunnels of their duffle-coat hoods. 'Who do you model yourselves on?' I asked, equally seriously. 'James Bond, mostly,' said an eager redhead; and a monumental 'fatty' said, with all the weight of the law, 'We have found some signs, and I think I know who did it. It was Jo Carpenter.' 'Is that a boy?' I asked. 'No, it's a girl.'

As I was leaving the playground the small Samantha came running across. 'There's a rhyme I forgot to tell you,' she said:

Scatty Bear
Lost his lair,
In the field
Of buttercups and daisies.

'I made it up myself,' she said, and skipped away contentedly.

Thursday 7 December

Cold, damp, and windy. Winter has now come fairly, and we need the aid of Father Christmas and all the angels of heaven.

Little girls are infinitely cosy, even at the age of seven. Just inside the playground one little girl greeted another: 'Hello, we'll play Tickling. But we won't play Tickling till Mandy comes, and Jayne—OK?' A programme had been arranged, an intimacy confirmed. Tickling, they said, was 'Just tickling people round their necks, and you chase them.'

A solitary Christmas wait stood, with her toes turned in, playing a plaintive tune on her recorder. It was not a Christmas tune, actually, she said. It was 'James Galway'; and she could not play it properly, Jennifer could. 'Jennifer learnt it off of the television, when James Galway played it, and she learnt it me.'

A rope had been tied to the fence and lay on the damp ground while the skippers finished eating their crisps. It was made of 'One piece of rope, one piece of ribbon from a dress, and one pyjama cord'. Someone picked up the end, and someone else announced 'It's Keep the Kettle Boiling.' They ran through the slack rope, not letting it touch them. A girl by the fence, holding a ginger-haired rag doll, called in a querulous voice, 'Please may I play, Helen?' No one answered her.

Two girls were playing the new version of Cat's Got the Measles in an angle of the wall. Two girls and four boys were playing Piggy in the Middle ('only it's got two piggies in the middle'). Nobody was playing marbles.

A third of the football area was taken up by a tremendous game of Red Rover. No statement of the rules of this game can convey its seething wildness, fierce as a border raid. 'You say, "Red Rover, Red Rover, we call someone over", and if they break the line they can take a player and if they don't they join the other side,' explained a bystander. The game has a primitive excitement; it produces a half-hope that, after all, fighting may have been the most intense way of living for most men—the hour of glorious life, not the sordid carnage.

In the quiet western part of the playground two small girls were practising signals from *The Brownie Guide Handbook*. 'Um—*T*,' said one, then, 'Is your arm supposed to be straight out?' The flat-faced girl, always happy to amuse herself on her own, was practising with a pea shooter. 'You've stepped on my pea,' she accused me (though I am sure I did not). 'I'll have to get another one.' A boy had the same kind of pea shooter, in a superior plastic. 'But they're not pea shooters,' he said, 'they're Popper Point pens, and you cut the nib bit off.' Also, presumably, you remove the tube of ink. One is never given full instructions for anything, by a child. It seems like laziness, but it may be diffidence. They wait for one to enquire about the next step.

The Rabelaisian pair came staggering up. One asked, 'You know why fairies have got sore lips? 'Cos they bin to the goblin' feast.' The other burbled, 'Hey diddle diddle, The cat done a piddle, The cow done a poo on the floor . . .' ('The tongue can no man tame. It is an unruly evil.' Perhaps they were possessed. They move constantly as they talk, as if possessed; whereas the 'intellectual' boy, who had arrived from the other direction, always stands in a poised and somewhat detached manner.) 'There's the one about the three little pigs, too. The three little pigs went into a pub, you see. The first one had two beers, and the second one had two beers, but the third little pig asked for six. "Why do you want all that beer?" said the barman, and the little pig said, "'Cos I'm the little pig that went 'Wee wee wee all the way home'."'

The intellectual nodded, straight faced. 'I know that one,' he said. 'Here's another one. There was a ghost with large green eyes. Up came another ghost with large green eyes and said, "I am the ghost with large green eyes." "You're going to be the ghost with large black eyes if you don't go away," said the first ghost.' 'Oh yes, *very* funny,' said one of the naughty boys. 'Why aren't you laughing then?' said the intellectual coldly.

I waited with the 7-year-olds by the gate and went out of the playground at the tail-end of their line. Their young teacher stood in the middle of the road, elegant in her high-heeled boots and pencil-slim skirt, dangling a large red crêpe-paper Father Christmas from two fingers. 'Look, there goes our Father Christmas,' said the boy ahead of me. '*Our* Father Christmas'—his own classroom saint, as it were.

Tuesday 12 *December*

Schizophrenic weather. Torrential rain earlier, then sudden brilliant blue. Rain clouds trying to gain the mastery again. A temporary duck pond has appeared in the pub car park.

Four comrades came striding out of school, two boys and two girls, saying, 'Shall we do the same as yesterday?' One of the girls stood in the centre of the netball court, and the others tried to run past her. The two boys ran past easily; she let them go. She intimidated the girl, while the boys prodded and teased her from behind. When the first tenseness of the game had subsided I asked what they called it. 'It's Semicircles,' they said.

A boy stood in the attitude of a Greek athlete, looking for pursuers round the corner of the wall, poised for flight. 'Did you ever hear of a game called Semicircles?' I asked. 'No,' he said, and ran.

There are always people inspecting puddles, on a day like this. They walk through the puddles from end to end, watching their toes and the movement of the water. One day they will think they know enough about puddles, unless they are poets. I watched a boy testing his car in the water, trying it as an amphibian, and wondered whether males incline towards objective experiments, females towards subjective. Whatever the answer, the experiments were personal and private, and I knew a question such as 'Do you love puddles?' would be unthinkable.

The same tripartite rope was in operation by the front fence, with four bouncy little girls skipping in it—so bouncy they jumped as they waited for their turn. No other skipping to be seen.

A dwarfish boy in shabby long trousers and an almost threadbare duffle coat came to report, '*He's* got bubblegum in the playground, that boy in the orange mackintosh, and we're not allowed bubblegum in the playground.' The next minute it was, 'Can you do up my shoe-lace, it's come undone?' Later there was a scraped finger to be pitied: 'Soon the blood will come—look!' Heaven help the woman he marries.

The naughty boys were not together this morning. The freckle-faced horror had a different boy in tow, and said, 'I want to introduce you to "The Dog of Liverpool". He's a sex maniac. He takes nude ladies to bed with him.' The Dog looked embarrassed, so I took an interest in his Deep Water Fisherman's badge. 'Did you write all that down? His name is Jonathan Price. What's his name? What's your name? Mary Jane. Where d'you live? Down the drain. What's your shop? Fizzy pop. Boom, boom!' He kicks in the air and exclaims 'Boom, boom!' whenever he makes a joke, having picked up the habit from the television puppet Basil Brush.

I found myself face to face with a female centaur, the front half of whom was protecting herself with a red flowered umbrella. 'What did Robin say when Batman got run over?' she asked, and sang the answer to the Batman tune, 'Ner ner ner ner ner ner ner ner—flat man!' 'I've got one, too,' said the back half, looking up with difficulty. 'What is black when you buy it, red when you use it, and grey when you've finished with it? Coal.' It was a pleasing riddle of the old pictorial kind, and appropriate for the Christmas season.

Other riddlers waited, wanting to say that the little chimney was too young to smoke, and the little telephone too young to be engaged. But I was more interested in 'Why does my child never bite 'is nails any more?' 'It is because,' said the microscopic pretend-parent, 'I knocked all 'is teeth out.'

The wind rose dramatically. A girl used her bright blue patterned umbrella as a sail, and collapsed with laughter when it turned inside out. Their mothers presumably send the girls to school with umbrellas because it is raining, but they find it is more fun to use them as toys.

Tuesday 19 *December*

A day brilliant as a Christmas bauble; the gold light was so dazzling I could hardly see the road. A small boy was pointing out his Christmas hopes in the window of the general store, and as his granny led him away she said in a soothing voice, 'Yes, you'll be busy, Christmas Day, won't you, with all your toys.' The grocer came staggering along the pavement with a heavy box of groceries. He nodded at an acquaintance, who remarked on the rich clusters of holly berries on the tree outside his little villa. 'Yes, my Mum planted that with a holly pip,' he said. School breaks up tomorrow.

The 7-year-olds came across the road and began skidding straight away. 'Want to join in?' I asked their teacher. She shook her head silently and hurried in to have her coffee. Half the fun of skidding is the falling down at the end of the slide, which they do with as much dramatic effect as possible, kicking their legs in the air. They make sure they have a witness: 'Dennis, out of the way—now *watch*. I found a good place, didn't I?' Hand prints appeared where they had put their bare hands in falling. 'Oh,' they said, noticing, 'it's where you put your hand,' and they made a few more prints in the frost. 'But you can never see the skids, can you?' The boys waiting their turn told me, 'When we came to school it wasn't so good. It's really good now. It's better than skating.'

A few people were skipping in single ropes. 'But it's really too cold,' they said. On the sunny side of the playground, which is also the football area, football games were still being arranged and delineated. 'Is that your coat?' said an organizer, dropping the garment on to one of the piles marking the goal posts, and the owner of the coat took his place as goalkeeper. They were smallish boys, though tough. One of the two who were picking up sides took a step towards the row of applicants lined up at the fence and snarled, 'Let *'im* play;' then stepped back and instantly became a sweet-faced little boy. The other chooser advanced on the line saying, 'I pick Derek;' then hooked Derek round the neck and led him lovingly to his place.

Boys by the wall teased the girls at long distance, chanting, 'Her, her, her,' and apparently wanting to be chased. I never learned whether 'Her' had any significance, and the boys later insisted that it was the girls who were teasing them. Neither could I get any sense out of a boy who was squatting on his hunkers by the wall, with his head zipped up to the eyebrows inside his jacket.

I was pleased when a senior boy with a grubby, earnest face volunteered, 'I got a dirty poem. You put people's names in it, what you know like each other—like Linda and Andrew. It goes,

> Down in the meadow where the green grass grows,
> There stands Linda without any clothes.
> Along come Andrew clipperty-clop,
> Down with 'is trousers, out with 'is cock.'

(One of his friends turned and shooed away two 8-year-old girls who had crept up. 'Go away, you shouldn't be listening, you're too young.')

> Two months later, all went well,
> Four months later, belly began to swell.
> Nine months later, belly went pop,
> Out come a baby—['Ooh!' said one of his audience]
> with a nine-inch cock.

It was recited in all seriousness: not to shock, but as information in which I was known to be interested. He would no more have smirked during the telling than would Miss Dean-Smith the folksong scholar have done while discussing one of the bawdy songs she studied.

Some of the older girls had been standing beside us, waiting for us to finish. 'Everybody's frozen,' they said. 'Yes. Is anybody doing anything except sliding and football?' 'We got our disco this afternoon—for the fourth-years.' 'What do you do at a disco? I know you dance—but what else?' 'Food,' said the tank girl. 'You have to bring your own records,' the other girl went on. 'It starts at 4 p.m., after school, and it finishes at 6.' 'But what dances do you do?' 'Disco dances, like the twist. And you can get a girl's hands and swing her through under your legs and pull her back again. You can learn it off the television. You can watch *Top of the Pops*. Or you can see it on that film *Grease*.'[1]

A few somewhat inanimate boys had attached themselves to the group. (It is not true that children become more animated in cold weather, though some have the sense to hurl themselves into sliding or football. Most of the people in the playground were like myself; they were inclined to stand in the sun and revive their freezing extremities by blowing and stamping.) One of them said stolidly, 'What's big and green and salt? The Incredible Sulk.' Then, because I do not watch a television programme called *The Incredible Hulk* at 7.30 p.m. on Saturdays, they patiently explained that it is about a man who is put together piece by piece ('It takes four hours') and when he gets angry 'his eyes go all funny'. 'And he is called the Hulk,' said the boy. 'And he *does* sulk.'

The 8-year-olds, neatly lined up in two lines, did not seem at all Christmassy. No talk of Christmas presents hoped for, or visits to 'my Nan'. Perhaps they were too occupied with keeping warm. The girls were jigging about with their hands in their pockets. The boys were swiping at each other with scarves. Nevertheless, when I murmured that I needed something about Christmas before I left, a boy said, 'OK, here's a rhyme,' and supervised me while I wrote down:

> We three kings of Leicester Square,
> Selling ladies' underwear,
> Not very safe to wear—

He faltered, and said uncertainly, 'No elastic in them.' Another boy said scornfully, 'It's not like that, 'e's got the words wrong. It's like this:

> We three kings of Leicester Square,
> Trying to sell ladies' underwear,
> How fantastic, no elastic,
> Trying to sell ladies' underwear.'

'Are you sure the last line is right?' 'Yes, of course, but it hasn't finished yet. There's a chorus.

[1] Paramount Pictures' liberating teenage film of 1978, starring John Travolta and Olivia Newton-John.

O star of wonder, star of light,
The royal knickers caught alight,
How fantastic, no elastic,
Guide me to the traffic lights.'

'Well, thank you,' I said, 'I'll have that for my Christmas present. 'Bye now, and have a happy Christmas.' They all smiled and waved as they went into school.

Postscript. When I first heard that parody of 'We three kings of Orient are', in 1952, it was making fun of the 'spivs' who sold, at high prices, goods that were in short supply after the war. 'We are three spivs of Trafalgar Square, Flogging nylons tuppence a pair, All fully fashioned, all off the ration, Sold in Trafalgar Square.'

Tuesday 23 *January* 1979

'The time is out of joint.' This is the beginning of the third week of term, but the first time I have been able to come to the playground. My mother had a stroke in Capri and I went out there for nearly a fortnight. Yesterday the council workers went on strike for more pay; today the train drivers are on strike for the same reason; and the lorry drivers have been on strike for weeks.

The village street was covered by a transparent layer of melting snow, the consistency of Walls' Sno-fruit, through which the road could be seen clearly.[1] The pavements were still a pristine white, sullied later by wet yellow gravel flung from a council gritting lorry.

I met some of the senior children at the school gate, carrying cardboard boxes full of crisps. 'Can you see us across the road?' asked the tank girl. I saw them across to the old Infant School building. 'Aren't you coming out to play?' I asked them. 'No.' 'What a pity, I thought you'd be sliding.' 'No, we're not allowed. Too many people got wet.' They hurried into the school.

[1] A water-ice in a triangular cardboard tube, costing 1*d*, sold by Walls' mobile 'Stop Me And Buy One' ice-cream sellers in the 1930s.

Back on the opposite pavement I met the school secretary. 'What weather!' she said, 'I must go and get the milk 'cos otherwise we shall get no coffee and I'll have my head chopped off.'

A teacher came towards the gate. 'You've heard there's no playtime, have you?' she said. 'Yes, the children told me, thank you. I expect you had to close yesterday, didn't you, because of the council workers' strike?' (The withdrawal of the caretaker's services means the school has to close.) 'Yes, and I think several schools are closed today, because of the train strike.' I let her go in for her coffee.

Tuesday 30 January

Raw. (The opinion in the village was, 'It would be different if we could have some sun, but this is miserable.') The playground was covered with a mixture of snow, ice, and water.

A few children emerged doubtfully, and stood zipping up their anoraks and pulling on their gloves. 'Did you come out yesterday?' I asked them, 'I wasn't sure whether to come.' 'No, we didn't. It was snowing and raining. The snow first and then the rain. People are helping to clear it up today.' Some boys were indeed jabbing at the ice with spades and trundling the debris to the side of the playground. 'So what did you do instead?' A middle-sized boy said, in the rhythmic speech that is natural to children, 'I've been playing with cars in my classroom. Some people play Blow Football, and some people draw pictures, and some people make paper planes.' A tall girl, with medieval bobbed hair and far-away grey eyes, came and stood silently by my side. 'You didn't come out to play yesterday,' I said, for want of a better remark. 'No,' she said, looking tragically into the distance, and then, 'Would you like a nut?'

A boy careered round the corner and all but knocked me over. He was wearing brand new bright yellow gloves with strips of bright blue pitted plastic down the fingers. 'They're goalkeeper's gloves. They're for knocking the ball away. And they're sort of nice inside, sort of bouncy bits [foam rubber], and outside it is grippy.' He dashed away, and a boy beside me commented in a deep adult voice, 'But actually at the moment we are playing Cops and Robbers.'

A Robin Goodfellow of a boy flung himself in front of me and demanded, 'See that boy over there? Write down that he's a nincompoop.' The deep voice beside me said, 'Nincompoop means "blockhead'. I looked it up in a dictionary.' The other children joined forces in defence of the absent nincompoop; or rather, they defended him in the customary juvenile manner, by attacking his assailant. 'And write down that *he*'s always being told off in class.' This was a blow for a blow, but Robin Goodfellow was delighted to continue the exchange. His flattish pink face sparkled with merriment. 'He's an idiot,' he declaimed. 'He's got false teeth, and—and he's a big 'ead.' The others came back with, 'And *he*'s a twit, *and* he's greedy, *and* he's queer, and he's—he's an unidentified old-age pensioner!' Invention having run out, they pursued the boy with pointing fingers. They circled round each other, clowning, clouting each other in high good humour. 'He's the Wonder Woman, he's the Incredible Sulk, he's going nuts, he's going stupid, he likes kissing girls—*no*, he likes kissing *boys*.'

No one seemed to be playing any real (or 'structured') games. The children were getting used to a somewhat changed environment after their lengthy confinement. One group of older boys was sliding with dedication and skill, on a long and probably carefully selected strip of icy tarmac. It was a far cry from the random helter-skelter of the pre-Christmas sliding, when ice was a novelty. The boys were now taking turns and watching each other closely, with serious faces.

A girl was sitting on the steps, looking pensively at the ground. It is so unusual to see a child entirely on her own in the playground that I wondered if I should find out the reason. By the time I reached her she was surrounded by friends and was saying to one of them, 'You *did*, you know.' I hope they are used to my eavesdropping; if I allow myself to think of it in ordinary terms I know it is rude. The solitary girl kindly explained what the trouble had been. 'Jill was meant to come to tea, and she forgot. But I'm going to sleep with her soon. No, there's nothing wrong at home. We often go and sleep with each other.' Nothing remained to be said. I could not think of any small talk except that it was a pity the ground was too messy for skipping or any other games. 'We've been playing chasing, sometimes,' she said. 'A *kind* of chasing. If you're caught *there*, you hold there, and if you're caught *there*, you hold there, and gradually you can't move. It's called Operation Chase.'

Robin Goodfellow, still up to his merry pranks, ran past shouting, 'You're just a lot of old benders.' The girls said, '"Benders" is what he calls old women,' and they shouted back, 'Martin sucks his thumb and curls his hair at the same time.' It is true that he has beautiful curly hair, like a faun.

Some of my old acquaintances gathered outside the senior classrooms, waiting for the teachers to arrive. 'Do you want to know how to make an Irishman dizzy?' they asked. 'Tell him to pee in the corner of a dustbin.' 'And do you want to know what to do if your nose is on strike? Picket.' 'You *are* up to date,' I said, and they nodded sagely. 'We've got some more,' they said. 'How do you make an Irishman burn himself? Phone him when he's ironing. And did you know that Father Christmas is Irish?' 'No.' 'Well he is. Although there is a front door and a back door, *he* comes down the chimney.'

The girls went in, and two boys who had been queuing up stayed to tell their jokes. The smaller boy, in a red and white woolly beanie hat, began carefully, not looking at me: 'There was this boy and a priest. And the boy sat on his lap—he was a *little* boy—and the priest says, "Where do you think heaven is?" and 'e says, "In the toilet," and 'e says, "Why do you think 'in the toilet'?" and 'e says, "Every time my Dad wants to go to the toilet 'e says, 'Oh God, are you still in there?' "' He looked at me with quiet satisfaction, having completed the story and got it right (I know the feeling well). 'That's a good story,' I said, 'where did you learn it?' 'I was on an Adventure Holiday,' he said, 'and some boys told it me then.' 'Thank you,' I said, and turned to the other boy. 'What was the joke you wanted to tell?' I asked. 'I didn't want to tell one, I was just listening,' he said.

Tuesday 6 February

The hoar frost had half melted on the grass beside the playground. I walked round reading the sodden paper litter while I waited for the children to come out: a Kit Kat wrapper seemed old fashioned and offered no entertainment; a Smiths Football Crazy Crisps bag would have appealed to a football fan; and a wrapper from a Trebor Toffee Flavour Phantom Chew, no bigger than a cigarette card, provided two pictures of ghosts—one white, one green—as well as some invigorating letterpress.

A dark-fringed, pink-cheeked little girl approached me, playing on her recorder and at the same time holding out her bag of crisps so that I could take one. She had the wholesome ordinariness of a Red Ridinghood in a nineteenth-century picture book. The crisps were more than mere crisps; they were bats, and came from one of the three varieties of Smiths Horror Bags. 'They have bats, bones, and claws,' she said. 'These are bats and they have Batburger Flavour. I don't like the bones, but I like the other two.' Smiths offer the currently required '*more* for your money'; not only fantasy, but a sophistication of shape and flavour that creates specialized tastes.

By now we were surrounded by girls playing 'The Lord of the Dance' on their recorders, standing in a half circle like waits left over from Christmas. I suppose boys learn the recorder, but they do not linger lovingly over tunes, as the girls do. Red Ridinghood said she did not know 'The Lord of the Dance'. 'They learn by singing the song and then playing it. They can find out the notes for themselves. I can't do that. I have to watch them, and see what notes they play. I learn like that. Shall I play you a bit of "Silent Night"?' 'That's all I know so far,' she said, finishing. Most people are honest, and modest, and she was 'most people' personified. 'Oh, there's my mother and my granny!' she exclaimed with joy—'Goodbye!'; and she ran to the fence.

Strings of girls and strings of boys were running about the playground. The girls said they were trains; the boys said,

'We're doing the same, but we're bomb trains because we bomb into people.' The mouse boy, at the end of a five-boy train, took it upon himself to explain (he had offered to show me, 'slowly'; then he relented when I said, 'I've got to write down what you do—couldn't you tell me in words?'). 'What happens is, Richard pulls us along and swings us round and then Clive lets go of me and I bomb into someone. *Now* we'll show you.' They went off saying, 'Whoo-whoo.'

Red Ridinghood returned to my side saying, 'Look what granny gave me—"Chipsticks".' ('Chipsticks' are, naturally, another form of potato crisp.)

The boys playing football were making more noise than I had ever heard on the playground before. I was assailed by several of them as soon as I ventured near the game. 'Have you got Football down? Let's have a look. Yeh, F for Football.' They were wildly excited, and in a baiting mood. They said about twelve people were playing and it was six-a-side, but it looked like an unholy scrimmage to me. They kept going off and coming back to report the score, although the game was going on without them and I could not see any scoring done. 'Now it's 2 : 1.' 'Now it's 2 all.' 'Now it's 10 all.' I tried to bring the conversation back to reality. 'How do you score?' 'You get the ball in the goal, don't you? It goes in the net, don't it?' 'Yes, but you haven't got a goal here.' 'Do you know how to play netball?' 'Yes, you throw the ball through one of those rings.' 'That ring,' said one of them, tapping my wedding ring. 'The ball wouldn't go through there—he he he he! What's that you are wearing on your ear?' The conversation was getting personal, and I became a little dignified. I was conscious of standing with my back to the wall, being attacked by half a dozen dervishes. They started pushing each other about. 'Have you got fighting down?' asked a bystander. 'Have you got ear-ring down?' came an echo from the midst of the scrum.

Children play innumerable nonce-games—played today and forgotten tomorrow. Close by, a starved-looking, almost hatchet-faced girl was screaming thinly, 'Let me out of here, let me out of here!' Her hands were tied together with a home-knitted multicoloured scarf, and she was being led about by a girl whose face was expressionless in the frame of her gabardine hood. 'I am her slave,' shrieked the starved one; and I thought how, forty years hence, people will say, 'She looks as if she has had a hard life.' Scarves are much used as games adjuncts, and mothers are wise to make them from odd balls of wool. In winter they bandage eyes in the time-honoured game of 'leading about and guessing where you are'; in summer they are brought to school to tie legs together for practising the three-legged race.

Perhaps the most noticeable game was the simple throwing and catching of bright new tennis balls between two people, sometimes extended to a circle with penalties for dropping the ball. No other game could be said to be 'in'. Some girls were playing leap-frog, but so casually they could not get the frog to bend down properly—she was more interested in eating her crisps.

Boys lying on their stomachs on the ground were playing a boxed game called *Frustration* ('Fast and Furious Pop-o-matic Family Fun', it said on the box). The dice was shaken by thumping a transparent plastic dome in the centre. In spite of the space-age embellishments, the game looked uncommonly like Ludo.

The bell-ringer walked by, and the children groaned. They admitted, though, that they were 'not really playing anything these days'. The little girls rising from a camp by the fence had not played anything: 'We were going to play Round the World. We just thought of it when the bell went. Someone's on it, you know, and they spin you round and say you land that you could be a ballet dancer, you *are* a ballet dancer. Then they say, "Turn the lights on," and you start being whatever you are. And when they say, "Lights off," you've got to go down in a little ball on the ground, and they touch you all except one, and you go round that person singing, "Wakey wakey sleepy head, Don't forget to make your bed." It's not a Brownie game—oh, *no*.' It had slowly been dawning on me that I had heard of this game before, but I did not like to interrupt. Last September the girls called it Eggs and Bacon.

I met the school secretary rushing out of school in her usual forthright and jolly manner. 'Next week is half-term,' she said, 'Next week that ever is—so don't come, will you?'

Wednesday 21 *February*

Savagely damp. The surface of the playground was an abstract pattern of damp patches.

A playground is a bleak place without children in it. I was glad to hear the 7-year-olds debouching from school and lining up in their playground over the road. The boys shouted, 'Eh—eh—eh—eh—eh!' as they came running out and crashed against the child at the end of the line, shunting a ripple down the whole line. A teacher appeared, a refugee from the weather in shapeless brown trousers and shapeless brown anorak. She moved the children briskly across the road, saying as they passed, 'What are you doing, Alastair? I've got my beady eye on you.' (He was, she said later, 'Running and trying to overtake the child in front'.)

Four of the girls immediately started playing Follow the Leader (thus named) and tripped daintily after each other. The leader was constantly turning her head to see that the others were following, and giggling. They were as self-conscious as if they had been asked to set the game up for television; which perhaps shows that they do not play this game often enough for it to come naturally.

Two little girls began dodging round me, the larger one bouncing at the smaller one as frighteningly as the monster puppy in *Alice* and saying, 'Let's—play—IT!' They were off before I (being a serious and enquiring person, like Alice) could ask whether 'It' was now the usual name for chasing. Some girls who were huddled in duffle coats, watching, thought it was an allowable name. 'It's a name everyone knows, but we normally call it just Chasing,' they said (which is what I had thought).

We stood and surveyed the mêlée. It was difficult to distinguish a single real game. One tends to forget that the playground is the arena for the whole gamut of the children's social life, of which games are only a part. Three tiny girls in a niche by the front door were simply gossiping. A girl who looked as if she was playing 'It' turned out to be fleeing from retribution. She was being pursued by a fourth-year girl who, it seemed, she had insulted. The fourth-year, a tall 11-year-old, forced her to her knees and said sternly, 'You apologise. Go on.' The second-year looked up at her and bleated 'I'm sorry.' 'Get up. *Up*,' ordered the older girl, and set her free.

I was accosted by a mature fellow who introduced me to his friend and said, 'Us two and Venables, who is really Mark Reynolds, are running a newspaper.' 'Venables?' 'It's from a Jennings book. I've got two. I'm Jennings and this boy is Darbishire. We're making this magazine called *The Form Three Times*—we're not really Form Three, it's all in the book. I've done an Italian holiday report—my brother went there. And, um—' ('And, um,' mimicked one of the girls who had been eavesdropping.) 'And—what was it?' ('And—what was it?' mimicked the girl.) He dived at her, scragged her, and went on, 'Oh yes, I did a report on the rail strike.' It is good to know that Anthony Buckeridge's pleasantly old-fashioned school stories are still being read.

The children and I trailed towards the classrooms together when the bell rang. A boy with a large sandy head of hair assured me it was 'Too wet and cold for marbles', and he added, 'There's not much holes left, either. They've mended all along there by the fence.' He cheered up and said, 'Can I tell you a joke? What is the cure for water on the brain? A tap on the head.' Other boys came prancing around, to make a clowns' festival. 'Why did the zebra cross over the zebra crossing? Because he was a zebra.' 'Knock, knock, who's there? Medam. Medam hand got stuck in the door.' 'Knock, knock, who's there? Doctor. Doctor who? You just said it.' (That is still the most popular of the 'Knock, knock' jokes, and has probably existed since soon after the *Doctor Who* television series began on 23 November 1963.) 'How do you send some elephants on holiday? Pack their trunks and put them on a Jumbo jet.' 'Wheee!' 'What's green and barks? A grape Dane.' The sandy-headed boy suddenly rounded on one of the other kids and said fiercely, '*He* threw a football in Daniel's ear.' The accused boy was much taken aback and said nothing. 'Was it deliberate?' I asked. The sandy-headed boy considered the question, with his head on one side. 'I don't know,' he said.

Most of them went in. I was left with a few boys and a girl

who had been pleading, '*Please* can I tell you a limerick?' 'OK,' I said. She recited:

> There was an old lady from Wheeling
>   ['Wheeling?' 'Yes, Wheeling.']
> Who had a peculiar feeling.
> She sat on a chair,
> With her legs in the air,
> And did it all over the ceiling.

'You'd all better go in now,' I said.

Friday 2 *March*

Dank and raw and overcast: but the birds were singing.

The footballers were out first, roaring at each other and setting up their games. Most of the footballers' utterances are animal yells indicating joy or displeasure, or incoherent shouts. The spectators are the only ones who can be understood as they remark, ''E got that easily,' ''E's good,' or 'Martin's slow, en' 'e?' At the edge some more formal players were picking up sides, only one girl among them. On her heels I read the label 'Superkids'. 'Are those football shoes?' I asked her in passing. 'No, not really. They're trainers,' she said.

I was approached by the boy who reports for *The Form Three Times*. He looks more professional than I do, he wears his pencil behind his ear. 'Can I have something for *my* notebook, for a change?' he said. I began giving him a short version of the Opie publicity piece. He did not seem very interested, and soon rushed off to interview a girl about a large soft plastic spider, saying, 'Excuse me a moment, I've got to get this down.'

The tank girl's thin-faced friend was with someone new today. She arrived with her arm round the new friend's waist and 'doing a three-legged race, only we're not tied'. The new friend said helplessly, 'She says it's a jogging game. We have to sing "*Da*, da *da*, da *da* da da *da* da da," and I know the tune but I can't think what it is.' She was happy to join in a revel that had nothing to recommend it except lack of meaning. Some children, rather than play known games, are constantly inventing worthless occupations of their own, and can apparently always find someone to join them.

Paul, the sophisticate, came ambling up and said, 'I've got a *sweet* little story for you. It's about three bears who went up in an aeroplane. There was Daddy Bear, Mummy Bear, and Baby Bear. The plane caught fire. Daddy Bear said, "There are only two parachutes—we'll have to leave Baby Bear behind." So they jumped out of the plane, and when they reached the ground there was this little squeaky voice said, "Me not daft, me not silly, me hold on to Daddy's willy."' It really was a sweet little story.

The children were idling, this morning. Apart from the footballers, who are always serious and whole-hearted, they were behaving in the careless, lethargic way in which the adult world has been behaving since Christmas. Three large untidy games of Donkey were being played against the high wall; and there were two nondescript ball games further along—one of throwing and catching, the other of kicking a ball against the wall.

Suddenly some small boys began jumping up and down, tremendously excited, outside the entrance to the girls' toilets. 'He's run into the girls' loos!' they chorused. I never saw the boy who did it, but he was undoubtedly a hero for the rest of the day.

After the bell rang the children and I gravitated towards the classrooms, telling jokes on the way. 'What's tall and wobbles? The Trifle Tower.' 'What's black, white, and red, and can't move in a lift? A nun with a javelin through her head? 'Doctor, doctor, I've only got fifty-nine seconds to live! Wait a minute.' Two girls wanted to know, 'Why is Egypt like a frying pan?' and were prompt with the explanation, 'Because it's got Greece at the bottom of it.' 'But it hasn't,' I said, puzzled. 'Well, we don't know. That's the riddle,' they said, and we stood, defeated.

Nobody could think of another joke. A boy with a round rosy face and a fringe, a very wholesome helpful boy, said, 'Write down that I'm brilliant.' 'Ah yes, that reminds me, what words do you say for something that is good? I know "brilliant" is one.' 'Yes, that's the main one. And, like, if you're not feeling well, you can say "I don't feel too brilliant." Or if something's good you can say it's "ace" or "fantastic" or "skill".' 'What about things you *don't* like?' That was more difficult. Mostly they thought of food they hated and they made faces and said 'Yuck', 'Grue', 'Oo-argh'. 'If you want to explode, you can say, "*Chicken* feed!"' suggested the boy. 'Say she drops a bottle of milk,' he said, waving a hand at a girl, 'She could say, "*Chicken* feed!", or she could say, "Oh, *Dracula*!" And I expect you've heard of "damn", haven't you?'

The boys had to go, and the girl was left in charge of me; a nice girl, rather motherly, who wore her dark hair in two bunches and said she wanted to be an archaeologist when she grew up. She obligingly tried to carry on where the boy had left off. 'What goes put-put-put-put?' she said shyly. I could not think what it could be (I usually look up in the sky when I am pretending to think, so that they are not prevented from remembering the answer). 'A bad golfer,' she said, pleased with the outcome. 'I really must go now.'

Thursday 15 *March*

A maelstrom of wind and small wet snow. Trying to be colloquial, I said to a friend in the village street, 'I don't expect the kids will be out in this lot'—but they were.

Not many, but there they were, all boys: running and hiding, sliding and yelling, in hoods and scarves and zipped-up arctic jackets. No one thought of playing football, and the empty football area seemed twice its usual size. One girl came out, and stood pulling up her socks.

A boy paused briefly, wanting to tell some jokes but not willing to wait long. 'I should think you'll like these,' he said eagerly. 'Why would you take a pencil to bed?' I was not going to waste his time. 'Don't know,' I said. 'To draw the curtains,' he said triumphantly. 'And why would you lie on the edge of the bed? So that you could soon drop off.' He beamed, and ran away.

Two more girls came out, cautiously. 'Isn't it *cold*,' said one. 'We haven't been out all week. We've been staying in with wet breaks all the time. We've been reading comics in our classroom.' Her companion looked at her accusingly: 'We haven't got any comics in our classroom.' The first girl laughed and said to me, 'We have. At the back. She doesn't know—she's new.' The new girl changed the subject. 'I hate this day,' she said, 'it's my brother's birthday. He's 7. I hate my brother, he's horrible. He's pretty tough. I hit him and then I get told off.' She was such a sweet-looking girl, too, with a peaches-and-cream complexion and angelic golden hair. 'I expect he hits you first and you hit back. Is that what happens?' I said soothingly. 'No, it doesn't,' she said, with Freudian vigour.

'Not many girls out, are there?' I said. 'About five,' they said. 'Round the other side. They're chasing. And those boys are playing Om Pom. The girls don't come out so much. Some of them get in a bad temper with the cold.' The new girl said, 'My mother gets in a bad temper even without the cold.' They looked around for games. 'I think those boys over there are playing British Bulldog [they were, oblivious of the snow] and there's lots more girls out, look.'

The new girl began to suffer from the cold, vociferously. 'I'm getting frozen,' she said, 'this is stupid.' The other girl said 'I really came out for some fresh air.' 'Yes,' said the new girl, changing tack, 'the others are sissies. Oh gawd, we've got maths next, and my hands are freezing!' 'It's meant to be spring next week, isn't it?' said the other sedately. But all the new girl said was, 'Ow, my face!'

Even stoics will retire if honour permits. The imbiber of fresh air called to Mr —— as he came down the steps, 'Is it the end of break?' ''Cos we haven't got a bell today,' she explained to me carefully, ''cos there aren't many people out, so we ask a teacher.' Mr —— approached, a friendly gnome-like creature, his bearded face framed in a snow-fringed duffle hood. 'Not quite yet,' he said; and then, to me, 'Do you go to other playgrounds? The games vary, don't they, from place to place? For instance they don't seem to do the hand-clapping they used to in Swindon, my previous school. I can't remember the rhymes.' I looked fondly at 'my' children. 'You know clapping games like "Under the bram bushes"[1] and "A sailor went to

[1] See I. and P. Opie, *The Singing Game*, 453–5.

sea", don't you?' I said. 'Oh *yes*,' they said (the new girl too). Mr —— said, '*I* never see them,' and accusingly, to the children, 'You never do them, do you?' 'Yes we do,' they giggled at him, 'in the loo.' Honour was satisfied all round. Mr —— said to me happily, 'Well, you'll have to go inside the loo then,' and to the children, 'That's the end of playtime. Off you go!'

At the front of the school I found the lingering end of a chasing game. A boy was teasing the chaser and, while he jeered, took up the traditional teasing stance, with hands lightly on wide-apart knees, ready for a quick get-away. The boys hailed me as a friend in a bleak landscape. 'We're playing Stuck in the Mud,' they said. 'But what was he shouting?' I asked, longing to write down the archaic-sounding rhythmic defiance. Nobody could remember, least of all the boy who had just yelled the words. They all tried to think. 'At the beginning of a game you can shout, "Turn round, touch the ground, bagsy not It", and then you needn't do the chasing,' said one. 'We sometimes say "Can't catch me for a toffee flea!"' said another, 'but most of the time we just tease the person who's coming after us. We call him "Jam sandwich", and things like that.' Behind the milling boys I could see two small girls placidly skipping in an orange woollen scarf tied to the railings. Then two young woman teachers came briskly out of school and without any palaver lined the 7-year-olds up and marched them over the road.

Wednesday 28 *March*

It had looked beautifully springlike through the window; sunny, and the daffodils nearly out. Outside, a furious, infuriating wind nearly skinned me alive.

The marbles season has definitely begun, by fair means or foul (I feel piqued that it seems to have been brought on by a ban on football). Children of both sexes raced to the marbles fence as soon as playtime started, carrying marbles in bags made of towelling or corduroy, and in tins and zipped pochettes. Two girls tried to start a game by the gate, but were still complaining that the wind was blowing their hair in their eyes, and smoothing it ineffectually with their hands, when I left them to find more action.

About a month ago a boy, explaining why no one was playing marbles, said that 'they' had filled up all the holes. Now, whether through rain, excavation, or serendipity, there are plenty of holes along the foot of 'the marbles fence'. The complete length of the fence was crammed with players; players who, although they might have hunched shoulders and one hand in a pocket, were dedicated to their game. 'Isn't it too cold for marbles?' I asked one boy. 'Oh *no*,' he said, but I noticed he was playing in gloves, all the same. The skilful players have not lost their skill during the winter. I saw a boy propel a glass-fourer into a hole from at least fifteen feet, and his friend said, 'Oh yes, 'e's a good player. 'E booted one from right out of that drain yesterday and it went right in.' The drain must have been twenty feet away from the hole.

Behind the row of leaping, shouting players ('I'm second, I just baggsied it!') strolled the riff-raff who collect at any sporting event: the tricksters and entertainers. 'Want a joke?' I was asked. (I had almost expected it to be, 'Want a tip for the 3.30'?) 'There was this elephant and a little ant, and the ant bit the elephant's toenail, and the elephant said, "Why don't you pick on someone your own size?"' An urgent urchin, glad the joke had been a short one, gabbled, 'What did the big leaf say to the little one? Hold on, I'm just about to blow off.' He wrapped an arm round a friend's shoulders, and repeated the deliciously naughty words 'blow off, blow off' into the friend's ear as they strolled away. 'Shall I tell you about the tin can that got stuck in a dustbin?' asked another boy, and then, as I poised my pencil for a long answer, 'I can't tell you—it's rubbish.'

On the edge of the crowd a boy walked up and down making music with his hands. He had just inherited, from a fourth-year boy and a million boys before *him*, Wordsworth's art of blowing 'mimic hootings to the silent owls'.

I went to the lee of the school for shelter and found several other people there. One girl had an old skipping rope which she had been given, she said, by her second cousins who used to live in Canada. It was a genuine old rope, I am sure, with straight wooden handles tipped with brass (not one of the 'Victorian' ropes made from old Lancashire weaving bobbins that are on sale in all the 'nostalgia' shops). A womanly little dumpling sat beside her 'best and oldest bear', called Edward, who was wearing a red and white spotted handkerchief 'to keep his ears warm' and a pair of makeshift trousers of red felt and safety-pins. 'Tomorrow I shall put him in his mackintosh,' she said, 'which is a nappy rubber with two holes in it.'

A girl in a peacock-blue overcoat was leaning forward and saying hotly to another girl, 'You can't *have* it. You can't *have* it.' I edged towards the group, trying to approach obliquely, not like a teacher about to sort out some trouble. There was no sign that they had been playing a game. I meandered up, and one of them, an old acquaintance, said, 'There was a man standing in a field with snow all round him. What time was it?' 'I can't think.' 'Wintertime.' 'Oh, *why* didn't I think of that?' 'There were two biscuits crossing the road,' said another. 'One of them got run over. What did the other one say?' 'Crumbs,' I said, meanly, it being my neighbour's little boy's favourite joke at the moment. Then I asked the girl in the peacock-blue overcoat 'What was it she couldn't have?' She squirmed a little, and the other girl answered. 'We were playing TV Programmes. The letters were FB. It was the *Flaxton Boys*, and I guessed it, and she said I couldn't have it.' 'Someone told her,' said the other. 'Someone standing next to her told her.' So the game had come to a halt.

Girls waiting for their teacher by the 8- and 9-year-olds' classrooms said they had been playing Building Up Bricks, but the wind had blown the rope about a bit. They said they still play French Skipping with elastic, at home. Some of them had been playing what I call Camels; they knew the game as Humpty Horses.

I was fetched by the tank girl and her friend to repay a debt of honour. I had promised to fetch their marbles from the lane after playtime. 'Can you fetch our marbles now?' said the tank girl. 'There's a little silvery one that belongs to Karina and . . .' I picked them up and handed them through the wire fence. They accepted a green and white china that did not belong to them. 'We might as well,' said the tank girl.

Wednesday 4 April

The sky veiled in high cloud; the sun showing through, sharp-cut like a silver penny. The air cold and fresh.

The owner of Edward Bear greeted me. This morning she had a plastic carrier bag full of dolls. She is a bonny child, with red cheeks and flyaway black hair. A slightly fey boy sidled up and pulled a grey furry mouse out of his pocket (not a real one, as it might have been seventy years ago). At least half the children bring their 'familiars' to school. We talked of marbles, which are still in season. 'I don't play,' said the girl. 'Oh *no*. I've got four hundred at home, all coloured fourers, and I don't want to lose 'em.' I wondered if children's policy with their marbles is reflected in their financial policy later in life. 'What games do you play, then?' I asked her. 'I never get anyone to play with me,' she said, with no trace of emotion. ''Cos there's a problem with me. I've got a trouble in my tummy. I've had it ever since I was a baby. I keep having accidents. They call me "flea bug" and "flea catcher" and things like that, and they think I stink—but I don't.' I said, 'You look all right to me,' and asked the boy why he had a handkerchief tied round his neck. (It was an ordinary white man's handkerchief.) 'Because that's where I keep my handkerchief,' he said. 'Would you like to see how I climb up the school and get tennis balls down from the gutter? I hook my fingers in the grating, see, and pull.' He showed me.

Two girls paused long enough to report, 'She and me, we're playing Nicking Coats. You see, the boys nick our coats and take them to the boys' toilets and we have to get them out again.' 'Are you allowed in the boys' toilets?' 'No,' they said, and went on with their game.

The caretaker was shouting, 'Come off, come off!' and waving his arms at the boys playing with toy cars at the edge of the grass patch. 'He wants the grass to grow,' said one of the boys. He suddenly looked me straight in the face and said, 'I'm scared of nobody. D'you know why? I strangled a boy yesterday—half-strangled him, anyway.' Later I saw him fending off two assailants. His face was pink with the effort of not showing his fear.

Close by me two small girls each held an arm and a leg of a midget of a boy, and were swinging him back and forth perilously near the ground. They told me it was called 'A leg and a wing'. I was delighted to find that it was accompanied by a rhyme which gave it form and charm, and probably indicated a certain antiquity. As they swung the boy they chanted

A leg and a wing
To see the king,
One, two, *three*,

and at *three* released him so that he flew forwards and landed on his feet.

'Proper' games were few and far between: only the marbles along the fence, and some boys using a large white marble as a football in the football area. I looked for the lines, circles, and nodules of children that indicate a game is in progress. About eight girls were standing one behind each other on a line of the netball court. They were disconcerted when I asked what they were playing. 'What we're doing *really*', said their spokeswoman, searching her mind for an orthodox equivalent, 'is playing Follow My Leader. Well—the one in front has to have some crisps and she does an action and they copy her and if she thinks they are good she gives them a crisp and it goes all down the line. We call it Keep Fit Classes. We just call it that—it's a made-up name.'

Out of the corner of my eye I had been watching six or seven of the older girls playing what looked like Om Pom. 'But it's not ordinary Om Pom,' they said. 'You start by drawing a snake on a person's back and saying, "Draw a snake upon your back, Which finger did *that*?" and you poke them and they turn round and you spread your fingers out and say they choose that middle finger on your second hand, if you are counting in tens that would be eighty. Then they turn back again and you hit them with your fists—not too hard—and say, "Fast or slow?" and they say which, and you do it again and ask, "Fives or tens?" and they choose again, and then you run away and they count in the way they've chosen and then they start looking for you.'

Having managed to get this high-powered description down

on paper, we relaxed. 'Have you heard?' they said, 'we're playing Bottle Tops nowadays. We don't play marbles only. It's because of Richard Knowles. His mum and dad have a pub and he brings us loads. You have to sort of flip them and get one over the edge of the other.' Richard came to see why we were talking about him. He was carrying under his arm a large plastic food box full of treasure—marbles and bottle tops jumbled up together. I have seen children flipping bottle tops in Corfu, and heard of them doing so in West Africa, Spain, and New York, but I have never seen or heard that they do it in Britain. As usual the game seemed to have arrived from nowhere in particular.

In the corner of the playground the caretaker was trying to subdue loose piles of stiff coloured paper left over from the term's artistic endeavours. 'I'm not allowed to burn it any more,' he said. 'Those houses over there complained about the smoke.' He was stuffing it into cardboard grocery boxes, and protesting, 'Hey, hey, hey, hey!' at the children who circled round taking what they wanted. 'Only two more days,' he said with satisfaction. 'Roll on the holidays!'

Wednesday 25 *April*

Summer term started on Monday. Playtime was five minutes late after a furious, icy shower. Water was still pouring off the corners of the roofs and bubbles sailing down the gutters as I walked from the car park to the school. The playground was a waterscape, which shone when the light broke through the clouds.

Two little girls were turning a very long rope. They moved it so that it slapped into a lake of water and watched with total absorption as the spray rose in tall fountains. The other girls became impatient (not having control of this new game) and shouted, 'Move it! move it!' When I left they had a line of perhaps ten girls steadily Building Up Bricks and others coming up and asking, 'Can I play?'

The water was the central event of playtime. Boys were

cautiously walking into puddles, watching with detached interest as the water level rose round their shoes. One boy walked his plastic Action Jack spaceman through a puddle. In the football area boys were kicking puddles at each other, 'because it's too wet for football', and scooping the water at each other with their hands, as if they were at the seaside.

Along the fence, however, marbles games were being arranged. 'Did you bring any marbles?' they called out to friends, swinging their own marbles bags from one finger. One of the girls who had not brought any marbles told me, 'Marbles is coming up. It's the marbles season about now. It's *always* the marbles season about now.' 'Yes,' I said, 'it does seem to start up again in the spring. But there is sometimes skipping as well.' We looked at the three long ropes we could see from where we stood. 'We used to skip,' said the girl, 'but we haven't really done any skipping for a long time.' I knew what she meant. There is no sense of excitement, urgency, or importance about skipping at the moment.

A skirmish of boys whirled by; one boy laughed and retreated as another hit him, and others were in attendance. The pursuer gave up and went away, and the pursued dissolved into a torrent of explanation. 'He's a spaceman—that boy who was hitting me. He keeps on hitting me, and he keeps going br-br-br—wobbling his lips with his fingers like this. 'Is name's Andrew and he comes from outer space and he wears Elton John goggles—that's the pop singer who plays the piano, *you* know. 'E's got a crew cut and long teeth. 'E's a funny chap, but 'e's popular—being a spaceman.' Children are sent into the playground as much to exercise their imaginations as their limbs.

They gathered round me in a chattering crowd. 'Want a joke? You heard about the Irishman who went to the toilet? He couldn't find the tap.' 'You heard about the Irishman who fired an arrow into the air? He missed.' 'What about the Irishman who went into the car wash? He was the one on the bike.' 'There was three hundred Irishmen drowned, trying to bump-start a submarine.' The little boy who gets everything wrong managed to get his joke in at last: 'What's black and white and goes upstairs and comes down red? A newspaper!' The others gave him short shrift. 'I don't get it,' they chorused, 'newspapers can't walk, anyway.'[1]

The pace of the joke-telling had died down. They began to look at the sky for inspiration. 'You know the one about making an Irishman confused, don't you? Put three shovels on the wall and tell him to take his pick.' One of the jokers suddenly transferred his attention to my part in the proceedings. 'Why do you do this?' he asked. 'Is it your hobby?' I was reminded of Stephen Spender's experience in the Fire Service during the war, when every extra-curricular activity—his poetry and another man's distribution of religious tracts—was explained and accepted as 'a hobby'.

'By the way,' said a boy I recognized as 'Jennings'. 'By the way, we've given up *The Form Three Times*. We weren't getting anywhere. We hadn't printed it or anything.' It is not only the London *Times*, apparently, that has production problems.

People drifted towards the classrooms. Two girls were holding tennis balls (one green, one yellow). They said they had been playing Donkey and Sevensies. 'I've been practising Sevensies in the holidays,' said one. As nicely as I could I remarked that two-balls in the south was nowhere near as skilful as it is in the north. 'I know,' said the girl, 'I used to live in Yorkshire. You don't see it like that down here, and people haven't got anyone to watch and copy.' 'No one to compete against, either,' I suggested. 'That's right,' she agreed.

Wednesday 2 *May*

The day before the General Election. Big unruly cumulus clouds in a blue sky. The wind icy.

A grandmother, just arrived by train from the market town, was loitering outside the school to see if her granddaughter had come out to play. 'I don't know whether she'd like it if I'm

[1] The venerable punning riddle 'What is black and white and read all over?' The answer, 'A newspaper', is no longer being handed down with care or, perhaps, comprehension. In November 1979 I heard the question put as 'What is black and red and read all over?'

here,' she said uncertainly. 'On the other hand if she *was* out, she might worry if I seemed not to notice.' She moved on, heavily, pulling a plastic-wrapped bundle in a shopping trolley.

The little boy who never gets anything quite right, whose name might be Derek, attached himself to me. 'Can I draw a figure of a saint in your notebook?' he asked, showing me the figure of 'The Saint' (Leslie Charteris's 'Saint' from his popular books of the 1930s, now on television) on the bonnet of his push-along car. (Toy cars were much in evidence this morning.) 'My book's really for jokes and games,' I said. 'Can I tell you a joke, then? It is a joke my sister thought up. She's five. Why did the orange stop running? Because he needed a toilet.' The tall, pink-faced adult boy sauntered up and looked at us benignly. He never seems to join in any games; he has a supervisory role in the playground. 'Jokes?' he said. 'What's red and goes, "Peep, peep!"? A strawberry in a traffic jam.' 'Ha, ha,' said Derek, after he had thought for a while. All around us little girls were hopping on their hunkers, and squeaking. 'We're rabbits,' they said, 'and we're begging for crisps.' They hopped away. Derek watched them fondly. 'Ha, ha, and off they go!' he said.

The senior girls are skipping now. The long ropes were being turned efficiently and with a sense of purpose. Some athletic 10-year-olds were Building Up Bricks over by the toilets, and in the front of the school others were playing Unders and Overs. Even so, it could not be said that skipping is in season yet, in the way marbles is. The marbles games were close-packed along the side and front fences. A friendly mother with a toddler in a pushchair was walking up and down the side lane, picking up marbles that had escaped through the fence. 'I often do this,' she said, bending down elegantly from spindly high-heels. 'It means so much to them, doesn't it? The trouble is, the lane goes up such a long way?' She turned the pushchair round and steered it back between the potholes; a fairy godmother lady, her face framed in a red fox-fur.

I remained among the marbles. There were some nice chinas in play; and some glass marbles more lustrous than I had ever seen before. I was allowed to hold one up to the light. 'It's like petrol in a puddle,' I said. 'Lovely. All purple and blue and pink and green.' 'It's called a "rainbow",' said the boy who had lent it. 'They cost 12p, and you can get them smaller for 6p.' He went back to his game; but first he knelt and carefully brushed the ground between his marble and the hole, as curlers brush the ice in the path of the curling stones. 'You have to do that,' he said, 'so that it goes in.'

The adult boy had followed me, and was standing observing the marbles play. 'It's Election Day tomorrow, isn't it?' I said to him. 'We're not having it in our school this year,' he replied. 'It's supposed to be in the village hall. At home we're voting for John Madeley. He's good.' 'Why is he better than Margaret Thatcher?' 'I'm not sure. I think it's because John is a man and Maggie is a woman, and we like men best. Mr—is voting for him, look. He's got a label stuck in his window.' I looked over the road at the headmaster's house, and sure enough there was an orange Liberal label in the window. 'When is a car not a car?' said the boy. 'The answer's "When it goes into a lay-by."' 'It would be better if it was "When it turns into a lay-by,"' I retorted, still slightly nettled about Margaret Thatcher. 'That's how it was in my joke book,' he said flatly. 'It was *goes*.'

The bell rang and the boys finished their games of marbles. One boy walked back to the classroom swinging his marbles bag round and round by its string. 'It goes round better the more marbles you have in it,' he said. 'It's just something to do, really.' He twisted the bag round and round and let the string untwist, saying, 'You can do that, too.' We skirted a knot of boys who were examining a KP Outer Space crisps packet. One of them was holding it and trying to read the conditions of a 'Name the Planet' competition in which 'a hundred bicycles must be won' while preventing the other boys from reading them. 'I'm going in for it, not you,' he said.

Paul came gambolling up. He had brought a friend who, he said, wanted to tell me a joke. 'He's a new boy this term,' he said. He put an affectionate arm round the boy and said, 'Tell her where you come from.' 'Walderslade, in Kent.' 'Tell her the one about "Behind the bush".' 'You've got to say, "Behind the bush," to everything,' Paul explained, and waited for the performance to begin. The boy from Kent, not looking at me, began, 'There was a bare lady.' 'You've got to say, "Behind the bush," ' Paul prompted. 'She took off her top clothes.' 'Behind the bush.' 'She took off her underwear.' 'Behind the bush.' 'Where are you?' 'Behind the bush.' Paul was enjoying himself hugely. 'Now tell her the monkey one,' he demanded. Perhaps the boy could not face another dialogue of that nature with an unknown grown-up. Instead he asked, 'What did God say to Joseph?' 'I don't know.' 'Come forth. But he came fifth and lost his beer money.' Both recitations were slightly imperfect, but I thanked him warmly all the same.

Wednesday 9 *May*

At last a warm, hospitable airstream from the south, replacing the Arctic air we had until yesterday. The swallows arrived two weeks ago, disappeared, and are now back again.

The whitebeam in the playground was holding miniature silver chalices against the blue sky. I doubt if any of the children noticed. Boys were hurling toy cars across the asphalt near the foot of the tree. Some girls were making themselves dizzy, whirling with their eyes shut and arms outstretched, their faces raised in apparent ecstasy. They were self-absorbed; they had no time for nature.

The mouse boy and his full-faced crony had got hold of an inoffensive-looking lad and were dragging him towards me. E's three today,' they announced. 'D'you want to see us giving him a typewriter?' They pushed him on to the ground. The heavier boy sat on his stomach and began jabbing his chest as if working a refractory office machine. The birthday boy turned his head from side to side, his soft pink skin and soft clean hair vulnerable on the asphalt. ' 'E's really nine,' said the boys. 'Shall we give 'im 'is bumps now?' They picked him up by his arms and legs and banged his coccyx on the ground, counting nine—'and one for bad luck'. 'Now let's all get on top of 'im,' shouted the mouse boy, intoxicated by the ritualized violence; but the birthday boy escaped and ran away.

The full-faced boy turned to me—almost turned *on* me. The joy of bullying had him by the throat. 'You know we saw a

Martian, don't you? It had a face like yours, *and* a green jumper and blue trousers.' He looked at me with bawdy eyes, waiting for the provocation to sink in. I was non-committal. 'I'll tell you a good joke,' he said. 'What's a fart? It's stinky air from between your legs. There was this man and he [I was unable to hear the next part, the boy was reeling about and his speech was slurred] and the man said "Piss off!"' Sexually precocious boys—for this resembled a sexual preamble—always seem to be well-fleshed, Falstaffian.

Skipping is in full spate. Three classic games in long ropes were to be seen from where we were standing. (One of the ropes was a thick working rope, such as is used in the Shrove Tuesday skipping on Scarborough foreshore.) Most of the girls were wearing their red and white checked cotton dresses; the effect was summery and relaxing. Even the younger girls were giving their whole attention to the progress of the game. The turners were turning expertly, the skippers running round promptly for their next turn. The seniors had invented a variation on 'Down the Mississippi'; they touched the ground as they said 'Down'. The juniors were playing 'Cross the Bible one by one, two by two, three by three' (the two in the rope cross over as they change places) instead of their usual Building Up Bricks. (Do they wonder where the name comes from? It was once a dance step called Cross the Buckle.)

There were too many single ropes to count. It is, anyway, difficult to count single ropes. Girls bring them out in case they want to skip, and may have them almost hidden as they stand chatting and eating crisps. I noticed that two of them were holding ropes with handles made from old bobbins, such as are sold in National Trust shops. 'Do they work all right?' I asked. 'Oh yes, they're good,' was the reply.

I counted six boys' games of marbles and three girls' games along the fence, and one game on a drain cover. At the end of the fence, where there should have been a marbles game, a boy was on his knees examining a crushed crisp under a magnifying glass. 'He's just being silly, Miss,' explained a friendly native.

The boys are playing football again, with the medium-sized footballs supplied by the school (they are about seven inches in diameter). I saw a good many tennis balls in play, and someone wielding a ball in a pair of tights. In fact recognizable games were everywhere, especially the games played in parallel lines like TV programmes.

A girl stood holding on to the fence while other girls came and tapped her back. I was not sure what game it was, and they kindly stopped playing to explain. 'It's Bobs, Dies, and Excuses. I got it from my friend and she got it from her sister who goes to the convent school. I'm on it, you see, and they knock on my back, and I say something like, "One bob, three dies, and two excuses"—you give them like an allowance of different things—and they run away and I run after them and they use their allowance. They can bob down, or say, "I'm dying," or something like, "My head's falling off"—that's an excuse. We'll do you an example.' Girls are nearly always gentle and helpful; but they are usually also obedient, and they had no time for the example because the bell went.

Near the classrooms we gossiped while we waited for the teachers to arrive. 'My gran's mad about Margaret Thatcher. She had her poster in the window.' 'Maggie Thatcher, Maggie Scratcher, Maggie Maggots!' 'Knock, knock, who's there?' 'Granny, granny, granny.' 'Knock, knock, who's there?' 'Aunt.' 'Aunt who?' 'Aunt you glad we got rid of the uncles?' The verbal delirium burned itself out, and a girl remarked soberly, 'My brother's always making smells. My mum says it's because he always has baked beans for supper. Oh, goodbye, here's my teacher.'

Monday 14 *May*

Summer at last. The village was in holiday mood. (If a hot sun is important to happiness, as it seems to be, then Apollo is still worshipped.) A blowzy young woman in a black evening dress and Indian sandals stood chatting to a friend and eating an ice cream cornet. As I passed I heard her say to her friend, 'Not too keen on having children myself, are you?'

A grandfatherly man on a bicycle skidded stylishly to a halt by the fence and said to the boys who were clinging there, 'Good morning, lads, how are you? Want some crisps? What sort do you like?' 'Wotcher!' said the boys indulgently. 'No, we don't want any crisps, thank you.' The man was silenced for a moment. Then he said intrepidly, 'Did you see the football? Portsmouth'll win next time.' 'Gar—they won't get anywhere near it,' jeered the boys. Having fulfilled some kind of social obligation, the man rode off.

I knew better than to discuss League football; but the boys know my role, and I did not have to think of an opening remark. 'Did you hear about the very high gate?' asked one. 'I'd better not tell you, you won't get over it.' 'Now what was that other one?' said his neighbour. 'Oh yes, did you hear about the quicksand? I'd better not tell you, it takes too long to sink in. And there was that stupid one about "I won't tell you about the butter, you might spread it." That's been going for a long time.' Probably jokes become stupid when they are too well known. The bright-eyed first boy went on, 'Do you know what Dracula's ship is called? The blood-vessel.' And his heavily freckled neighbour followed up: 'What are Dracula's favourite songs? "Teeth for Two" and "Fangs for the Memory".' '*I* told you that one,' said the first boy, furious. 'My uncle told me.'

A little girl who had been bursting to tell me a joke—'but it's a very long one'—came to meet me at the gate, with the freckled boy in attendance. She began. 'There was this boy lived in a little bungalow and it had three rooms.' '*Four* rooms,' said the boy, 'don't you remember, because there's the mother, and the father—' 'No, three rooms,' said the girl. '*Four* rooms, Mavis,' said the boy, 'I'll slap you round the ear 'ole.' 'There was this boy,' said Mavis, 'and he lived in a little bungalow that had three rooms—remember that, *three* rooms. He went into the first room—that was the kitchen—and his Mum was ironing and 'e said, "Mum, you're doing the ironing wrong," and she said, "Who's ironing, me or you?" and she slapped him round the face ['Round the ear 'ole,' said the boy, *sotto voce*] and 'e went into the lounge and his Dad was there reading the newspaper and 'e said, "Dad, you're reading the newspaper wrong, you got it upside down," and his Dad said, "Who's reading the paper, me or you?" and slapped him round the face. And he went into the bedroom and his sister was making the bed and 'e said, "Sister, you're making the bed wrong," and the sister said, "Who's making the bed, me or you?" and slapped him round the face.' The little girl looked at me with sweet schoolmistressly admonition and said, 'Now remember there's three rooms.' 'He went into the *fourth* room—' she said, 'and now you're supposed to say, "But there's only three rooms," and then I say, "Who's telling the joke, me or you?" and slap you round the face.' The freckled boy would not admit defeat. He came straight back with, 'Did you hear about the Irishman who had a hen? He let it drown. Oh no, I think it was a duck.' His confidence crumbled and he went away to watch the marbles.

The playground had a settled, summery look. Girls were sitting round the drain covers as if they were picnic tablecloths, playing with small dolls or chalking pictures. Two large but fluctuating groups were playing Fairies and Witches, jumping over the twizzled rope when they felt like it and staying out of range when they did not. 'Nobody's doing any real skipping,' I observed to a girl who happened to be standing near. 'Oh no, it's *much* too hot,' she said. 'But we've got a game we want to tell you about. It's called Booklets and someone's on it and the other two choose an author and think of a book the author wrote and—say I'm on it—they tell me the author and I have to guess the book that goes with him and if I get it right we have to swap ends there and back and when you reach the line you shout Booklets and whoever shouts it first is on it. Me and my friend made it up. We've had Enid Blyton and the Famous Five, and Roald Dahl and *Charlie and the Chocolate Factory*—I like that one very much—and Mary Dansby and her Puzzle Books.'

I caught up with the freckled boy who was watching the end of a marbles game. The bell had gone, and one of the players said to the other, 'No chicken's out.' '"Chicken's out"', explained the freckled boy, 'is what you say when the bell goes or if you're losing. If he says, "Chicken's out," he clears off with all his own marbles.'

I sat on the ledge with the boys. 'The girls say it is too hot to skip,' I said. 'Is it too hot to play football?' 'It might be for some people, but not for us. It's not such a fast game when it's hot, of course. You kick it more and you don't tackle so much 'cos it tires you out.' The teacher arrived and called out, 'All those not talking to Mrs Opie go in.' Three of the boys stayed. We exchanged conspiratorial glances, but we could not think of anything more to say. 'Thank you, that's all I want,' I said; they went in as other boys rushed out, swinging bulging plastic bags and exclaiming, 'We're going swimming!'

It was lovely to walk out of the playground slowly, in the shade, in a cotton dress; to look down at the wandering waste water from the drinking fountain, and be reminded of summers long ago.

Tuesday 22 *May*

Unsettled. The sky was an extravaganza of fast-moving clouds, changing shape and colour, clowning, racing, spitting.

I was early, or the children were late. I leant against a concrete post and listened to two old-age pensioners passing the time of day. 'I've got lots of odd jobs to do,' said the man contentedly. 'We can still get around, that's one thing.'

Then 'Flop, flop, flop' two boys ran out to the marbles fence and went straight down on one knee. 'Your turn, Tim, I think.' The game must have been continued from yesterday afternoon's break. The 10-year-olds (or perhaps they are now 11) seem too old to play marbles at all (except, of course, that marbles is really a man's game): they are tall, sophisticated, articulate. 'Shall we play Supershots, Nicholas?' called one spindleshanks to another, and explained, 'Only me and Nicholas do it. I made it up, 'cos once I was playing a game and I hit it from a long way away and got it in. So we tried right from the wall, and I got it about like that—about three inches from the hole.'

A smaller boy wanted to know if I knew how to make an Irishman feel stupid. 'You put 'im in a round room and tell 'im to sit in a corner. My brother told me that one.' It sounded like something I ought to report to the International Court of Justice.

My persecutors arrived: the mouse boy and his prized ally, the boy who—though pink with temerity—dares to be rude to grown-ups. 'Hey!' said the madman, fixing me with a challenging eye, 'shall I tell you a joke? You've got a brain. Shall I tell you another joke? You're a human being.' The mouse boy piped up, 'And your watch runs on gas and so do you.' I kept my dignity and said nothing. They tried another topic of conversation. 'That boy over there comes from Australia.' (He does; I asked a girl.) 'He came about two weeks ago.' The Australian boy was playing football with more than youthful concentration.

As soon as I set foot in the playground I had noticed new garden seats each side of the school door, and small conifers in concrete pots. The tallest senior girl told me, 'Well, you know we had a sponsored Spell [spelling test] and raised some money for swimming? Well, there was some money left over and Mr —— asked us for ideas to make the school more tidy and we had the idea to take up some of the turf round the corner and make a rockery and plant some trees so that the children don't play round there and tread all the grass down.' Her face was firm and righteous; she had joined the grown-ups. The naughty boy butted in with a gabbled story of a man who went up a mountain that was covered with spikes and kept saying, 'I can't wait to sit down at the top.' He nodded his head all the while to encourage me to start writing it down, and several times assured me, 'It's clean, really it is.' The girl looked at him pityingly and continued, 'We thought a small tuckshop would be a good idea, in the canteen, and maybe showers to get us warm again after swimming, because people have been complaining about the cold these last few weeks. But in the end we got these seats and plant pots.'

The naughty boy had not given up. 'D'you know the definition of a Stone Age bra? It is an over-your-shoulder-boulder-holder. D'you know the definition of a glass bra? A smash and grab.' '*I've* got a rude verse,' said a ginger-headed lad who had been lurking hopefully.

> Down in the valley of the green, green grass
> I saw an elephant sitting on his
> Ask no questions, tell no lies,
> I saw a Chinaman doin' up his
> Flies are a nuisance, bugs are worse,
> That's the end of the Chinese verse.

He was proud to give me this gem, and, anxious that I should write it down properly, supervised every word.

A crowd had gathered. Someone at the back said, 'John's got one rather like that but it's about Lulu,' and pushed John forward. John moved round beside me and dictated conscientiously:

Lulu had a baby,
His name was Sunny Jim,
She stuck it in the bath tub
To see if it could swim.
'E swam to the bottom,
'E swam to the top,
Lulu got excited
And pulled 'im by the
Cocktail, shampoo,
Lay 'im on the grass,
And if 'e gets excited
Smack 'im on the
Ask no questions
Tell no lies
'Ave you ever seen a Frenchman
Doin' up his
Flies are a nuisance,
Bugs are worse,
And that is the end of
This little verse.

Some girls arrived and asked breathlessly, 'Do you want to know how to play Bottle Tops? There are two people and they have a bottle top each. [These 'bottle tops' are the crown caps made of metal, with crimped edges, that are used to close beer and soft-drink bottles.] You flick them in turn until they meet and then with one finger you flick yours so it gets on top of the other person's and the one who gets it on top of the other person's wins the bottle top that's underneath. It's a craze. We do it all the time. Skipping might come back later, but it's Bottle Tops now.'

They stood and looked at me. 'Don't you ever run out of games to write about?' they said. 'Not really. They are always a bit different, or there are different people playing them, or there's a new craze.' 'Oh, I know what you mean. About two weeks ago we were playing Jacks in the toilets and we got bored of that and started to play Donkey and then a boy broke a window with a football so *all* ball games were stopped, which was *not* fair.' She looked round and saw the queues forming outside the classrooms and said, 'Oh, we'd better go in.'

By the gate a girl with a narrow, intelligent face, whom I used to know when she was about 8, asked me, 'How long are an elephant's legs?' and was pleased when I said I did not know that 'they are long enough to reach the ground'.

Monday 4 *June*

Overcast. The air sub-tropically moist. The adult population of the village found the humidity tiring, but the children seem unaffected.

A boy ran out and took cover behind a corner of the school wall. Another zoomed out making a noise that was certainly not human. Another walked jauntily by eating an apple with a bottle top; as he scooped small portions into his mouth he assured his friend, 'I *like* eating it like this. I'm using it as Chinese food.' Children seem born with one lobe of their brain steeped in fantasy, which dries out as they grow up.

The crowd along the marbles fence was bigger than ever. Boys were showing each other the stakes they had brought to school. One was examining the contents of a brown plastic pochette and considering whether it was worth playing for. I looked into the pochette too. The marbles were scratched and chipped. The owner said defensively, 'I've got 143 at home. I've got a big blue see-through, but I keep that at home safe.' The other boys chimed in like small-town business men boasting to their wives: 'I've got an orange see-through,' 'I've got 179 marbles,' 'I've got 257.' They always know exactly how many marbles they own.

Somebody's marble had shot through the fence and could not be retrieved. The boys were making fanciful suggestions about how to get it back, and trying to throw stones so that they would hit the marble from behind and drive it towards the fence again. They shouted that they needed a stick, or a spoon, and pushed against the fence so that it began to sag. 'You could get out of this school,' said an 8-year-old in an awestruck voice. 'Look, we've made a big gap.'

The sweet hackneyed strains of the 'London Bridge' tune told me that Cross the Bible was being skipped; and near the skippers two little girls were clapping hands and chanting:

> Under the bram bushes,
> Under the sea,
> Johnny broke a bottle
> And he blamed it on to me.
> I told my Momma,
> I told my Poppa,
> And Johnny got a whacking on his—
> Ooh ah, cha cha cha.[1]

'Come on,' said the intelligent girl from last week, who has a new friend. 'Come and see what we're playing. It's rather a silly game, but Cheryl likes it. We play it with crisps. One of us holds out a crisp and the other one has to run past and grab it as they run.' They showed me. It was fun. I would have liked to have tried it myself. Mrs —— came up, sipping coffee. 'Do you ever find any new games?' she asked. 'Yes, I've just found one,' I answered.

They are still enjoying the nameless, casual lifting game I have noticed over the past ten years and have thought of as too unimportant to be worth describing. Today it had a name, which must raise its status. 'It's called Run, Pick Up and Throw. You put your hands together in a hand-grip, and the other person does the same, and you run and sort of hook on to her, and put your weight on her, and she lifts you round.'

Hand-clapping was more noticeable than usual. Instead of being used as a time-filler while waiting for the teacher, it was being played all through playtime. Several times I heard little chants arising from corners: 'Who stole the bread from the baker's shop? Number One stole the bread from the baker's shop' and so on.[2] And skipping was not being ignored; there were at least three long ropes in motion.

[1] See I. and P. Opie, *The Singing Game*, 453–5. [2] Ibid. 448.

The most polished entertainment was a game called Miming, played with great earnestness by three 9-year-old girls. 'One person is on it, and you can have TV programmes or pop songs or books or theatre, and you've got to mime it. If it's TV you draw a square in the air. If it's radio you do a smaller square. If it's a book you put your hands together and open and shut them. And if it's a theatre you draw some curtains. Then you say how many words it is and you mime each word—well, you mime each bit of a word, because if it's Starsky and Hutch you mime Star and then Sky. Then the others have to guess.' This is the modern form of Charades, played by all ages at social gatherings and often simply called 'The Game'.

The girls in the 8-year-olds' queue were sorrowing over a multiple murder. 'Some baby chicks have just been squashed. We'll show you.' The chicks were probably blackbirds, not long out of the egg and now extremely flat. 'It was a girl called Caroline. She trod on them.' 'Stamped on them, did she?' I said idly. 'Not really stamped. She just trod on them, deliberately.' They demonstrated how Caroline had lowered her shoe on to the fledgelings, calmly and—for all I know—with the most merciful intentions.

The boys in the queue were grumbling. 'We're getting bored with marbles, and we can't play football because someone broke that window, look. Not last Friday but the Friday before. William Johnson did it. I could kill 'im.'

Perhaps it was only because the value of marbles had gone down that the naughtiest of the naughty boys suddenly gave me one. He put it in my hand and said, 'Here you are. Stick it in your stamp album. It's a china.' It was an attractive small white china marble. 'You can't buy them like that now,' I said, murmuring my thanks. 'It's a peppermint,' declared his crony, the next in line. 'They come up the drains,' said the next. The queue was moving, and the young teacher bringing up the rear said, 'Ah well!' and I replied, 'As you say.'

Tuesday 12 *June*

A brooding, head-stretching morning. The sky was covered with a sheet of grey cloud which now and again leaked incontinently. I found that a Musical Coffee Morning was going on in the hall of the old Infant School, to raise funds for the school band. A party of recorder-players had assembled by the main gate and were about to cross the road to play a French folksong. 'But playtime will be as usual,' said Mr ——.

Marbles had spread to the roadside fence. Two girls favoured me with a look through their see-through marbles. One was 'an aqua' and the other 'a different kind of blue, a dark blue'. A prematurely middle-aged girl with well-formed features and a sedate demeanour came to ask if I wanted a joke. It was 'What part of a chicken has the most feathers?' While I was thinking, the rude boy barged up and shouted, 'The outside!' 'He's told you now,' said the girl, and we looked with sorrow at the rude boy and his adherent the mouse boy. The two of them seem permanently drugged with silliness.

I went to watch a boy and girl who were sitting at a drain-cover playing Bottle Tops. The rude boy reeled after me. 'Look, they're playing football,' he burbled, and tried to snatch my pencil. I concentrated on reporting the game. It was the first game of Bottle Tops I had seen. The oblong drain-covers act as built-in games boards in the playground. The girl sat at one end, gracefully asymmetrical with her legs to one side. The boy sat at the opposite end, cross-legged, his brain thereby poised directly over his body. He was a jolly, bespectacled fellow, confident of victory. He settled himself comfortably, rummaged in his bag of bottle tops, and announced, 'I'll play a Guinness.' 'I'll play a Yellow,' said the girl, and selected a plain yellow bottle top. 'You said you didn't have many Yellows,' said the boy accusingly. 'I know,' she said, 'but Dominic gave me some this morning.'

They put their bottle tops down on the drain-cover, crimped edges uppermost, and took turns at flicking them forward till they touched each other. The boy then put the tip of his finger carefully under his bottle top, trying to get a good purchase, and flipped it over so that it rested Guinness-side up half-way over the girl's bottle top. 'Hooray!' he yelled, congratulating himself in the modern sportsman's style by shaking hands with himself above his head. 'I've got millions,' he boasted. He looked round the circle of spectators. 'I beat that boy there eleven times running, yesterday—didn't I, Mark?' 'How many have you got, actually?' I asked him. 'A hundred and ten,' he answered without hesitation. The girl did not know how many bottle tops she possessed, and did not care. She was already riffling through the bottle tops in her bag and saying, 'I'm going to play a John Courage next.'

Suddenly all the spectators deserted the game and ran to the fence. A magnificent tar-spreading lorry was crawling along the street in a cloud of odoriferous steam. It bore the inscription 'Southern Tar Distilleries' and was attended by a bevy of men in tangerine overalls, who shouted to each other across the street. No one but myself returned to the bottle tops contest.

The game was not so exciting without an audience. The boy murmured to himself, 'I think I'll use a lucky one.' A 'lucky one', apparently, is a bottle top that has won a number of games already. I noticed that when he did not want to take his turn he banged his palm on to the bottle top instead. The girl was not aware of the rule, and had not noticed his action. When he won yet again the girl seemed disinclined to go on playing. 'Where do the plain yellow and orange tops come from?' I asked her. She did not know. 'We get people giving them to us,' she said. 'Loads of people go to pubs and scramble for them.' (She must have had at the back of her mind the juvenile custom of 'Scrambles!', when sweets or other treasures are thrown up to be scrambled for.) She assured me that the people scrambling in pubs were fathers and older brothers; and the boy added, 'My brother's a waiter in a hotel. He gets *millions* for me.' (Afterwards I found out that plain crown caps in different colours can be bought for wine- and beer-making in Boots the Chemists.)

There was some disturbance amongst the smaller girls. The little coloured girl was in some kind of trouble, and was saying that she didn't know and she didn't mean it. One of them told me in a shocked voice, 'She's been talking about my neighbour. She calls her Mrs Mumble. It's not Mrs Mumble, it's Mrs *Rumble*. She's being very *rude*.'

Someone who seemed to know me held out a tiny balloon made from a larger one. 'Look,' he said. We explored the whole subject of balloons, in general and in particular. 'I got a whole boxful,' he said, 'a cardboard boxful. I buys 'em over the road there. I saves 'em for Christmas. I got a whole lot, too many to count. And when the Brownies comes in the summer we puts water in the balloons and throws 'em at 'em. And someone tips a bucket of water over 'em, and then they go down in the muck, like—the mud that the water made. We makes balls of plasticine, or cement, and sticks pins in 'em, and we bombards the Brownies with our crossbows. We got their leader once, in the leg.' 'But where do you get the cement from?' I asked, picking up a small practical problem to begin with. 'There's a building site. There was a man making a chimney. We goes up the ladder, see, and gets the cement.' I was bewildered at the thought of so much savagery. I tried to sort out his story. 'But where are you when you throw the water bombs at the Brownies? And where are *they*?' 'It's when they're sunbathing,' he said. 'We creeps along and climbs the conker tree. They're lying underneath, and we 'its 'em on the 'ead with the balloons.' 'But what Brownies are these?' He looked round at his friends for information. Someone said, 'They come from London, I think. They camp in the village hall.' That explains it. They are aliens.

A small (though presumably senior) girl came out of the main door ringing the bell laboriously. 'Have we really had a quarter of an hour's playtime?' I complained. 'Yes,' she said firmly, 'fifteen minutes.' A group of 8-year-olds stopped long enough on their way to the classrooms to toss me a few cheerful jokes: 'Why can't you play cards in a jungle? There are too many cheetahs. What kind of hens lay electric eggs? Battery hens. What is yellow and white and goes a hundred miles an hour on the railway line? A train driver's egg sandwich. Cheerio!'

Wednesday 20 *June*

A classic summer day: limitless blue sky, hot sun, and the village butcher looking doubtful when I asked for a joint of pork.

The children were putting up trestle tables in the playground. A blond boy who sometimes acts as my adviser said, 'It's Charity Week at school at the moment. Have you forgotten? So we have thousands of things to do all this week. There are stickers you can buy to stick on yourself—Save the Children and some other charities. And you can buy balloons, and people blow them up and they go bang, bang, bang! Yesterday lots of them popped. You can buy cakes, and drinks, and *sweets*. You can't buy crisps.'

A tough little urchin was blowing up a Save the Children balloon and letting it go on long snortling flights. He blew it up and held it out so that I could read the red lettering: 'Save the Children. Give a Child a Chance.' 'It means Vietnam and places like that where children are hurt.' He let the balloon go and it nearly flew over the wire fence. He fetched it and blew it up again. The blond boy looked at him dispassionately. 'He chews it when he blows it up,' he observed. 'It could go down your throat if you didn't hold it tight.' The urchin tilted his head back, muttered, 'Here it goes,' and launched it again.

The blond boy is something of a loner, a non-participant. He looked at the smaller boys dragging benches and tables into place and said with a dreadful weariness, 'They are setting up, like—an obstacle race. You have to walk up that plank and go under the table and along the benches and—see that blackboard with that drawing of a little man on it? you spray that little dildo with a water pistol.' 'Dildo?' I said, alarmed. 'That's what I call him, anyway,' he said.

An eager little girl wanted to show me a sideshow 'like a big cardboard box'. She led me to the other side of the playground where the present-day equivalent of an Aunt Sally was attracting a crowd. 'You describe it for me,' I said. 'Well—it's like a big cardboard box and there's a round circle in the middle of the box and someone goes behind and sticks their face through the hole and any person who wants to can pay some money and gets a sponge and puts it in the water and throws it and tries to hit the person's face.' A queue of girls were arguing furiously amongst themselves. I thought they were customers. One of them said, 'You're wrong, it was *Diane*, then *Samantha*, and *then* you.' The speaker took her place behind the apparatus, tucked a towel round her neck, and stuck her face through the hole. I said to the face, 'Do you pay to put your face through?' 'No,' she said. 'Me and Diane and Samantha and Emma are running this stall.' The customers were far more relaxed; but then, they had none of the strain of organization. My informant had left the chief excitement to the last. 'Later on in the week they'll be having all the teachers putting their faces through—*and* Mr —— [the headmaster].'

We wandered away. 'How many different sideshows are there? *Ump*teen?' I said. 'Well, not umpteen,' she answered seriously, 'but quite a lot.' ('How does she know how many umpteen is?' I thought indignantly. 'She's only about 8 years old.') Shortly after this we parted company. I went to watch a boy trying to take as many clothes pegs as possible off a clothes line stretched between two high-jump posts, in one minute, using only one hand. A stoutish red-headed boy was lounging against one of the posts; watching, and wiggling a green plastic snake. 'It depends how big your hand is,' he said scornfully. 'I'm not going to do it.'

The girls in charge of the coin-in-water stall told me the secrets of their trade. They had a yellow plastic bucket two-thirds filled with water, with a 5p piece at the bottom. 'You have to cover—or partly cover—the 5p with a 1p piece. It's 1p a go, and you can win 6p. We've made 45p so far this week. We've only had four people win, but you've got a fair chance. Most people start from straight above and it doesn't work. You've got to start from the side.'

The economics of most stalls seemed about right. The roll-a-penny boy said, 'You roll what you like, up to 2p. We made 39p the first day, and 42p yesterday.' It was a proper wooden Roll-a-Penny, so called by the boy, with two sloping runways for the coins to roll down, and a large chequered board with numbered squares indicating the amount to be won if the coin settled inside a square without touching a line. Girls with a second-hand stall were charging 5p for books, 1p for plastic wrist-watches, 10p for dolls, and 10p for handbags. The Hit-a-Pin bagatelle in the corner was causing more interest than any of the other entertainments, probably because it was encircled by cheering spectators. An excited crowd fills the role of barker.

The stall-holders were playing up to the crowd. 'D'you wanna go? Give it 'ere then.' The crowd set up a shout of 'Woo-oo-oo oh-oh-oh!' as each ball started its journey, rising to a crescendo as it fell into a hole or found its way into a pin-fenced enclosure. 'Only 2p for 10 Goes. Score 250 or More to Win 5p,' said the poster; and this may have promoted the game more than any other factor. When the bell rang the stall-holders ruefully handed Mrs —— 2p. 'Is that all you've made?' she said sympathetically. We'll have to change the rules then. We'll have only one winner a day.'

By the gate the first-years were showing each other (in some cases selling each other) their winnings and purchases. My little dark-haired girl had won a broken plastic stencil in the Lucky Raffle. 'You have a piece of paper to undo, and if it ends with a five or a nought you've won a prize.' An irate shopper was looking at a book someone else was holding, and saying, 'I asked her how much that was and she said 15p.' Miss —— walked briskly to the head of the queue and said in a clear voice, 'Come on, first-years. In two lines, please.' The fun of the fair was over for the time being.

Tuesday 26 *June*

The village was sodden and defeated after hours of rain. The long grasses in the waste patch behind the pub lay pell-mell, like spears abandoned on a battlefield. And the rain clouds continued to patrol overhead.

Three girls ran to the fence as though they intended to play a game, and then stood there with blank faces. Perhaps they were waiting, as I was, for something to happen. One of them began to draw her finger along the line of raindrops which hung from the wire, demolishing them with quiet intensity.

The tough redhead sauntered up (the boy who had criticized the clothes-peg show last week). He is square, laconic, and as poker-faced as a cowboy at the beginning of a Wild West film. He did not say anything; he simply looked at the girls and they ran away squealing. He was dangling an outsize rubber band, half an inch wide. 'I flick it at 'em. It hurts,' he said shortly.

His face was expressionless. I said, 'I'm supposed to be collecting stories this morning.' 'Oh?' he said, interested. 'What kind? Little—um—jokes?' 'Long stories if possible.' 'I got this one about Pink Panther,' he said. 'There was this Irishman and 'e goes off to the Job Centre and 'e asks for a job.' He flicked idly at the girls as they crowded round. 'And the man at the desk sez, "Would you like a job?" and 'e sez, "Yes, please."' He turned round to accept a crisp. 'And then 'e sez, "Would you like this job with animals? Please answer these questions about animals. First of all, where does an African elephant come from?" And 'e thinks, and 'e sez, "Africa." Then the man at the desk sez, "Where does an Australian kangaroo come from?" and 'e thinks and 'e sez, "Australia." So the man at the desk sez, "Where does the Pink Panther come from?" And the man thinks and thinks and 'e sez, "Durham." And the man at the desk sez, "Durham? Why Durham?" and the Irishman sez, "'Cos every time it's on the TV it goes "Dur-um, dur-um, dur-um."' 'That's rather clever,' I said. 'Yeh. I picked it off my brother and he got it from a friend at school—the comprehensive school.'

The raconteurs had gathered round and were waiting for a chance to perform. A refined-looking boy with a lock of dark hair falling over his forehead said eagerly, 'Has anyone told you that Kermit song?' He and a cherubic blond friend recited to me, line and line about:

> Kermit the Frog
> Was having a snog
> And messing about with Miss Piggy;
> He pulled down her knicks,
> And stuck out his dick,
> And now they have three little piggies.

'It's terrific, isn't it. We got it from the third-years,' they said blithely.

'I've got a quickie, I've got a quickie!' the redhead urged, as the bell rang. 'There was this little boy and he saw his Mum in the bath and he said to his Mum, "What's that?"' ('Oh, this one's good,' said another boy appreciatively.) 'And his Mum said, "It's a slot." And then he said, "What's that?" and his

Mum said, "It's a button."' ('Belly button, *you* know,' from the crowd.) 'And then he looked a bit higher and said, "What are those?" and she replied, "Two bells." And the boy said, "If I put a penny in the slot and push the button will the bells ring?"' The audience murmured approval, looked round and debated whether to go into school, and then decided there was time for one more story. 'Hang on, hang on, I'll tell 'er,' said the redhead. 'There was an Irishman went into a pub and said to the barman, "Could I have a pie, and could you cut it into four pieces?" And the barman said, "Would you like me to cut it into eight pieces?" And the Irishman said, "Sorry, I'm not all that hungry."' ('I don't get it,' said the refined boy. We all explained.) The boys saw that the playground was empty. 'We'd better go in,' they said, and they all left except the redhead. 'Hang on, hang on,' he said desperately, 'Here's another quickie. "Kermit the Frog—"' 'I've had that one,' I interrupted. 'Hang on,' he said, 'it's different. "There was Kermit the Frog, Was having a snog, In the swamp by the river. He had it off with Miss Piggy, Gonzo, and Fozzy. But the one he liked best was the Swedish Chef."' I suppose it was different, in a way.

Thursday 5 *July*

A young man was cutting the long grass behind the pub with a fag hook. He was stripped to the waist, and the sweat was running down his chestnut hair and beard. The goat which used to keep the grass in check left with its owners when the pub changed hands.

Little girls in red and white check dresses were sitting on the ground. Some were playing Football Marbles ('which is a much better game than Bottle Tops') on the oblong drain-covers, and some were just sitting. Everybody agreed that it was 'boiling hot'. A boy (an impish, active-looking boy, with huge pop-eyes like a Palmer Cox Brownie[1]) said, 'I'm not playing *any* games at

[1] The Brownies drawn by the American author-illustrator Palmer Cox in the 1880s and 1890s were extremely popular—so popular that Kodak called their children's camera *The Brownie*.

the moment. I just sit or lie down.' All the same, he remained standing while telling a string of riddles. 'Why did the policeman cry? Because he couldn't take his Panda to bed with him. Who invented the five-day week? Robinson Crusoe—he had all his work done by Friday. How can you catch a squirrel? Climb a tree and act like a nut . . .'

We were interrupted by a jubilant procession of boys escorting someone who carried a square plastic cake box. (Fame is generously celebrated in the playground. When anyone does, or possesses, anything noteworthy, the children make him the centre of a procession. They make hooting, whooping noises to announce his arrival, and dance round him; and the stragglers who bring up the rear explain to the populace what an interesting person he is, and why.) 'He's got Wood Ants. They eat wood. They're huge. They're called Army Ants. Well, I think they're called Army Ants, because they're big and they might bite.' There was a discussion about whether the ants bite or not, and if so whether it hurt, and if so, for how long. I was given a long, close look at the ants, which had mostly taken refuge in a hole in a half-brick they had been given as furnishing. The boys tried to poke them out with blades of grass. 'They are like ants I used to know in Dorset when I was a child,' I said. 'They're beautiful. Big red heads and shiny black bodies. And they do bite.'

We relaxed into a company of about ten, who did not know what to do next but did not mind. I said, 'Somebody's sucking something that smells, and I can't think what the smell is.' 'It's me,' said the ant-showman, 'I've got a cough-sweet.' 'Have you got a cough?' I asked, because one never knows. '*I* haven't heard you cough,' said another boy, and turning to me said scornfully, 'He's *always* sucking something. Usually when it's a hot day he sucks fruit sweets.' 'And when I get home I have an ice lolly,' said the other placidly, not a whit abashed. The pop-eyed imp-boy managed to slip in another riddle: 'If slippers are slippers, and shoes are shoes, what are boots?' 'Boots,' I said. 'No, chemists!' he chortled, and ran away.

Three of the older girls arrived, wanting to tell me the news that they were playing Spaceships. It was real news, it was happening that moment and they could only spare a minute to explain what was happening. 'There are all different planets like Crispland, and Drinkland, and Feastland, and we travel there and when we get there we do what it says—like in Crispland you eat crisps, and Feastland you gobble up the food. And some people get stranded on different planets, and they start talking a different language. Say I'm on Pluto where the dogs are. Well, I take a spaceship and go to the other planets and make them speak English.' They went zooming off again.

A rather worried little boy wanted to know if I was coming tomorrow. 'I want to say that one they say to you if your shirt's hanging out, and I can't remember it properly and I might by tomorrow.' 'Does it go "Dicky, Dicky Dout, Your shirt's hanging out"?' I asked, and he nodded, delighted. 'Do you know one about when it's raining?' he asked. 'It's raining, it's pouring?' 'That's right,' he said, and recited,

It's raining, it's pouring,
The old man's snoring,
He went to bed,
And bumped his head,
And couldn't get up in the morning.
The doctor came,
And pulled the chain
And out came
An aeroplane.

'Or "Out came a choo-choo",' said one of the boys who had gathered to hear the performance. 'Or "Out came a [inaudible]",' said the rude boy, and the mouse boy tittered admiringly. 'They're stupid boys, they are,' said the performer. 'Mouse is not so bad. It's the other one.' 'Mouse? is that his nickname?' I said, thinking he must have seen my notebook. 'Yes, it's his nickname. He looks like a mouse, don't you think?'

The good boys went in to school, and I was left with the two 'stupid boys' and another who is nearly as bad. The rudest said, 'I'll tell you a poem. You put somebody's name in. It goes,

Going down the highway,
The speed of twenty-four,
Stephen did a blow-off
And blew me out the door.

'Where did that come from?' I asked. 'I made it up,' said Mouse. '*You didn't*,' said the rude boy. The third boy stood behind Mouse and prodded Mouse's head ceremonially with his forefinger while intoning, 'Mouse made it up. He did. I know. Mouse made it up.' Having testified to Mouse's honesty, he ran into school. The rude boy looked after him and said sourly, 'He's fat. He wobbles faster than I can run.'

Wednesday 12 *September*

Term began last week, on Tuesday 4th. After days of pure calm summer, the weather is hostile: overcast, with a creeping wind, threatening something worse.

The old Infant School playground is filled with 7-year-olds (a bumper crop this September, it seems, to match the glut of apples, pears, nuts, tomatoes, and runner beans). They came over the road to be greeted by their older siblings. 'Eh, you slow coach,' said one older brother unfairly, swinging his marbles bag. He said to me, 'I wait for 'im to come over 'cos 'e's quite new.' I could have paired these brothers and sisters to win a competition; they were startlingly alike.

The area round the marbles fence was already alive with players. I could not see much else going on. One boy was holding up his stake, a large glass marble, between thumb and forefinger, and lounging against the fence while he looked for a likely opponent. 'Robert,' he said, 'got any colour-fourers?' Robert hadn't. The boy said he was waiting till he found someone with a colour-fourer to play against, 'because my marble is kind of interesting'. He held it up to the light and turned it round slowly, admiring it.

The short boy (full of libido, a typical half-pint) was pulling at a strip of caps, exploding them one by one. 'Don't you burn your fingers?' I said, ever the solicitous female. 'Well, you do, but it doesn't hurt. What you do is, you pull the caps off like this and you breathe the smoke in and puff it out again, like you're smoking. It tastes just like a cigarette.' The small comedian had arrived, and confirmed the tale. 'You can get a cap bomb, and chuck it up in the air. It comes down, it goes bang, you pick it up, you put it in your mouth, you breathe in the smoke, and it comes out of your nose—just like a dragon.' He was rapping out the instructions like a sergeant-major. It was very convincing. 'But doesn't it give you a sore throat?' 'No, not really. Well, it gives you a sore throat that lasts for a day or two. You can't swallow. If you go to school with a sore throat then you don't have to eat a lot of the school dinners.'

Two girls came round the corner, shining with excitement. 'There's a glove gone up in the gutter,' they said. 'We were playing catch with it.' They were as excited as adults are when there is a flu epidemic, or somebody else's drains are blocked. I was not going to get involved. 'All you want is a tall boy, and a smaller boy to go on his shoulders,' I said decisively. A heroic look came over the face of the short boy, and he took charge of the situation.

I went away, untrammelled. I passed two little boys and, overhearing one say to the other, 'I got a hundred now, a hundred football cards,' remembered the magic of a hundred, or a thousand, or a million.

There must have been a leak from the toilets. A snake of water was pushing its way across the playground attended by several delighted boys who were encouraging it and making themselves into bridges for it to run under.

The glove had been rescued, and the small hero caught up with me. He showed me a brown burn mark on his finger. 'Poor finger!' he said in mock commiseration, and hit it. 'You got this rhyme?' he went on.

> Mary had a little lamb,
> She also had a duck.
> She put them on the mantelpiece
> To see if they would fuck.

Some taller boys arrived. 'I got a joke,' said one of them, 'but it might go on for ever.' I poised my pencil, hoping to write down a long joke before the bell rang. 'There were three men called Pardon, Pardon-Pardon, and Pardon-Pardon-Pardon,' he began. 'Pardon-Pardon and Pardon-Pardon-Pardon fell over the cliff. Who was left?' 'Pardon,' I said obediently. 'There were three men called Pardon, Pardon-Pardon, and Pardon-Pardon-Pardon,' he said. 'Pardon-Pardon and Pardon-Pardon-Pardon fell over a cliff—' We all began to giggle. 'Who told you that?' I asked him. 'My cousin,' he said. Cousins are necessary for a rich social life.

There was a ripple of interest as a boy came marching towards us with staring eyes and arms held stiffly forward. 'Look out! look out! he's a robot!' cried his escort, clearing a path. In the robot's wake came the mouse boy, his arm round the shoulders of a new friend. 'What happened to that rather silly friend you had last term?' I asked him, feeling I had a right to be outspoken after the rudenesses I had suffered. 'He moved to Devon,' said the mouse boy. 'What did the big candle say to the little candle? Are you going out tonight?'

The bell rang and in the queues the children complained, 'Nothing's happening. Only loads of tests to see if we've remembered anything after the holidays. One girl, Samantha Page, she took the First Years' Maths Test and she got *nought*—and she's 9.' They were thrilled with this scandal and went into school in high spirits.

At the gate Mrs —— was addressing her long line of 7-year-old newcomers. 'When the bell goes, then it's time to finish. So it's nice and quiet now, please, all the way across the road.' A small boy at the end of the line quickly clinched an argument. 'That is *mine*,' he said, removing a marble from his friend's grasp.

Wednesday 19 *September*

A cold front had been pushing through, wet-breasted, all morning; but at 10.35 the rain was so small it was merely dampness, so the children came out to play.

I was surrounded, almost captured, as soon as I set foot inside the playground. 'I got a joke,' said a boy, gently bouncing with enthusiasm. 'If you get a peanut and it gets stuck in your ear how do you get it out? You put a chocolate in your ear and it comes out a Treet.' A Treet is a chocolate-covered peanut sweet, the subject of jokes polite and impolite for many years now.

'I've got a joke too,' said a bigger boy. 'Why do some bulls have bells? 'Cos their horns don't work.' 'I got one, a short one,' said a girl (the well-behaved girl who waits her turn). 'It's a book called *Falling Trees*, by Tim Burr.' 'And *Flower Pressing*, by I. Flatten'em,' said another. A boy on the edge of the group started laughing, shaded his eyes with his hand, looked into the distance and said, 'It's a bird! Or is it a plane?' He turned his head, pretended to spit, wiped his eye and said, 'It's a bird!'

'I got a funny one,' piped up the most junior boy. 'They're all meant to be funny,' said the bigger boy witheringly. 'Well this one's about a German,' said the small one. 'It goes, "What did the German policeman say to his chest? You're under a-vest." Arrest, a-vest, d'you see?'

'Can I tell you a story?' asked a girl with cropped sandy hair, who stood tall and steady in a multicoloured sleeveless pullover and sensible shoes—I had never met her before, I was sure. 'Yes, please do,' I replied. 'There was a man and he was a millionaire and 'e had a Rolls Royce and 'e was riding along a road one day and 'e came round a corner and 'e crashed into a motor-bike and 'e fell off a cliff and a monk came along and 'e picked him up and took him back to the monastery and said, "Is there anything you want?" and 'e said, "Water! water!" and so the monk said, "All right then, I'll get you some water, but you're not to touch that switch." So the monk went up and up and up— ['Why did he go up and up and up for the water?' I asked, echoes of Jack and Jill rolling round in my mind. 'I don't know,' she said.] Well, anyway, he got the water and—um—after a few days he was better again and so he was sent home and after a few days he had enough money to buy a Mini and 'e was coming round the same corner and 'e fell off the cliff again, so the monk came along and picked him up and took him back to the monastery and said, "Is there anything you want?" and 'e said, "Water! water!" and so the monk said, "All right then, I'll get you some water, but you're not to touch that switch." So the monk went up and up and up and got the water and after a few days he was better again and went home and this time he only had enough money to buy a motor-bike and so 'e was coming round this same corner and he crashed and 'e fell off the cliff again and so the monk came along and 'e picked him up and took him to the monastery and said, "Is there anything you want?" and 'e said, "Water! water!" and so the monk said, "All right then, I'll get you the water, but you're not to touch that switch." So the monk went up and up and up to get the water, and the man touched the switch and guess what happened.' 'I can't think.' 'The electric light came on.'

Everybody took a deep breath. 'That kind of story is always in a pattern of three, isn't it?' I said. 'Where did you get it from?' 'It's just a joke,' said the girl vaguely. 'I got it from my brother.'

The bell rang, and I could not believe playtime was over so soon. I complained to an 8-year-old boy in a queue. 'Surely that was a very short playtime?' 'No, I don't think so. I was playing marbles.' I complained to Mrs —— when she arrived. 'Was that a specially short playtime?' 'No, the usual time,' she said, looking at her watch. 'I should have protested if it wasn't. You must have been carried away.' 'Yes,' I said. 'I think I was.'

Wednesday 26 *September*

Warm and damp. The house-martins were circling low. A wind sprang up while I was waiting for the children to come out and the first fallen whitebeam leaves began somersaulting towards the fence and escaping through the mesh. The notice board outside the school gate is now up to date, with Keep Fit classes beginning tomorrow and the Save the Children Fair in the village hall on Saturday.

Suddenly the boys were tearing past on their way to the marbles pitches, and before I could get there were in the closing stages of their first game, bent double and showing ten inches of bare skin between trousers and T-shirt.

I was intercepted by the boy who used not to get anything quite right; he is now a head taller and possibly a head wiser. He said, 'Are you still collecting jokes? Do you know how to sink an Irish submarine? You knock on the hatch and they open it up. I'll go and get my friend. He knows a nice joke.' The friend (short tousled curls and wide-open blue eyes) left his game of marbles and began breathlessly, 'There was this man—it's a bit rude, so—and he had this camel and he tried to snog [i.e. kiss] its bum and 'e can't keep 'is 'ead still because the camel kept pushing 'im with its 'ead and the man 'e keeps falling over and 'e falls in a pond and 'e gets out eventually and 'e sees this beautiful Indian girl's 'ead stickin' out the sand and then 'e says, "If I get you out will you do anything for me?" and she says, "Anything in the world" so 'e digs 'er out and then 'e says, "Can you keep this camel's 'ead still for me?"' (The children tend to drop their 'h's when they are speaking fast and put them in when speaking more slowly.) The story had come out all-of-a-rush, and while he was telling it the boy twice dropped the marble he was fiddling with and retrieved it. He did not really look at me until he had finished his tale; then he said, 'I got it from Martin Lucas. I don't know where he got it from.'

We relaxed and watched the marbles. 'It's funny, some of the boys are standing and throwing the marbles,' I said. 'They're not bending down at all.' The first boy, who had taken upon himself the role of guide and mentor, said, 'I can tell you about marbles. It doesn't *have* to be jokes, does it? What they do is this, they just start by standing and throwing, and they get down and flick when they've got their marble near the hole, and the one that's nearest the hole has the first go.' 'And there's a green chalk arrow showing where the hole is,' I observed. 'That's just for fun,' he said, and went on, 'What do elephants play in the back of the car? Squash.'

His narrow face became thoughtful. 'I've got another joke,' he said. 'There was this woman lived in the middle of a big block of flats. Above her was a blind man and below her was a doctor. She was in the shower and she heard a ring at the door and she put her towel round her and went to the door and it was the doctor and he said, "Congratulate me, I've got a new job." So she went back to her shower and she heard another ring at the door and she thought, "I won't bother to put a towel round me because it'll be the blind man," so she opened the door and the blind man said "Congratulate me, I've just got my sight back."'

In the senior classroom queue the girls kindly reported on the games in vogue since, as I explained, I had been writing down two long jokes. 'We've been playing Chase,' they said, 'and some people have been throwing and catching with balls, and a lot of people have been playing Salt, Vinegar, Mustard, Pepper on the steps. There's somebody on it, and if they call "Sea" that means everyone has to run to the wall and back to the steps without being caught.' There was no time to hear the rest of the rules, but the game seemed to have much in common with a game on the steps called Hoppit Lands, which I saw being played on 16 May last year.

'Oh and there's a little clapping song. Come on, let's show her.'

> Under the lilacs she played her guitar,
> Played her guitar, played her guitar,
> Under the lilacs she played her guitar,
> Played her gui-tar-ar-ar-ar.

I said, 'It goes on about a man who says he loves her, and he didn't, and she dies, and he laughs—it's too long to tell you, but it's in the *Gilwell Camp Fire Book*. That song must be about eighty years old, at least.' 'Gosh!' they said.

Wednesday 3 *October*

The sun occasionally showed as a shining disc, but could not break through the thin cloud layer. Beneath the cloud the air was moist and warm.

Two footballs appeared first, kicked by a boy who must have been told to bring them out, for he himself ran to the marbles fence. On the other side of the playground there was an explosive clatter, like a burst of fireworks. A juvenile millionaire had thrown handfuls of bottle caps into the air from a large carrier bag and shouted, 'Scrambles!' By the time I arrived the caps had been picked up and were being used as skids by twenty or more successful scramblers. Someone called out, 'Scramble some more!' and the profligate scrambler threw up four more handfuls.

The girls gathered round me in a companionable way, eating crisps and not bothering to make conversation. One of them held up a crisp and said indignantly, 'This crisp is *cheesy*, and the packet is "Salt and Vinegar"—it says so, look!' I have often wondered whether they notice the difference between the different flavours. I should have known they do; they take their food as seriously as do the very old. After the crisp-tasting was over we were free to admire a charming little plastic doll in lacy tights called a 'Pippa', and a baby doll in a cradle, sucking its thumb.

Nearby three of the older girls were skipping in a heavy long-rope which moved slowly to suit their mood and the enervating weather. One skipper wore high-heeled shoes; one wore a stylish yellow thick-knit golfing jacket, and had one hand in her pocket as she turned the rope; and one held a half-empty packet of crisps while she went through the motions of 'Cowboy Joe from Mexico' with casual grace. 'OK, we'll do it this way now,' said the turner, and started turning the rope so that it came up towards the skipper instead of moving down over her head.

I looked beyond them at the effervescent footballers and said to the girls, 'There's nothing new, is there?' 'There is one game,' said a cheerful, chubby girl. 'We made it up. You need about three people and then you hold their hands and they are North, South, East, and West, and you spin the two people on the end round and you say, "North, South, East, and West," and then you point and if they stop where you're pointing that person is on it.'

The tall bespectacled boy walked by, blowing ear-splitting noises on a blade of grass. 'It makes an awful racket, doesn't it?' he murmured. 'My mother hates it.'

The smaller boys congregate on a bench under one of the whitebeam trees to watch the football. The whitebeam leaves were falling round them. 'Do you know any games with dead leaves?' I asked them. 'It's good luck if you catch one,' said the Palmer Cox Brownie. 'I was catching leaves while I was getting conkers, for a lucky term next term.' 'Oh,' I said, 'have conkers started? I haven't seen anyone playing.' They said, 'There's a boy over there with a conker,' and escorted me to him. The boy pulled his conker out of his pocket. It was all threaded ready, on black tape which was wound round and round his finger. He could not find an opponent, he said.

Some boys were clambering up the wooden fence and looking hopefully into the next-door garden. 'They're waiting for their ball,' I was told. 'It's an old lady lives there. She usually brings it back to them.' I could see the white plastic football lying on the far side of the old lady's grass plot. The boys grew desperate and swayed on top of the fence calling, 'Wakey, wakey!'

I walked back to the classrooms with several lads who had a yellow tennis ball. 'We played Stinger with it—you know, chucking it at somebody while they run across by the wall. Yeah—it's brilliant. Or you can throw the ball over the roof—the small sticky-out bit of roof—and there's a person on the other side to catch it. Do you want to watch us play it?' They began instantly, not waiting for an answer. If I was a teacher I would have each boy fitted with a brake.

They sobered down while they waited in the classroom queue, and stood dreaming and picking their noses. 'What do you do with the—um?' I asked. 'The bogies?' they said. 'I flick 'em, like this, off my thumb,' said one boldly. 'There's a trick,' said another. 'You put one finger up your nose, and pull it out, and then you quickly put a different finger in your mouth. It's a trick, you see, to make them think you've eaten the bogy.' Some boys at the back began to heave and splutter. 'If someone's showing off', they said, 'you tell them,

> Show off, show off,
> Pick your nose and blow off.'

The queues were disappearing into school when two of the older girls came running up. 'We've got a joke for you and we've got permission—cross my heart.' 'Cross your heart and hope to die?' I said. 'Cross my heart and hope to die, Stick a needle in my eye—*and* in my thigh,' said one fervently. 'The joke's about an Englishman, a Frenchman, and a Scotsman, and they were flying over this cow in an aeroplane, and the Englishman said, "That's an English cow, it's got English dust on its feet," and the Frenchman said, "Oh no, it eez a French cow, it has zee French dust on its feet," and the Scotsman said, "Oh no, you're both wrong, that is a Scottish cow, can't you see its bagpipes?"'

Tuesday 9 October

Still unseasonably warm, though less humid.

A plump little girl (a morsel fit for an ogre to eat) was squatting, knees apart, on the asphalt, while others stamped up to her from behind, chanting. It was a ritual I could not identify, and when there was a pause I asked them to explain what had been happening; seven 7-year-olds, helping each other to piece together the history of a game. 'We were playing—what's it called again?' 'Snow Children.' 'It's our own made-up thing, it's not off TV. Melanie made it up, didn't you Melanie?' Melanie awoke to her responsibilities with a start. 'We just have a travel in icy lands. Those people were stamping 'cos it's snow and you mustn't get your feet stuck. That one that was squatting? She was squatting 'cos she was drinking water, and the others were shouting, "Don't drink it, it might be poisonous." It was out of a stream. But it wasn't poisonous, because otherwise we'd be dead, wouldn't we?' A scornful voice said from somewhere behind me, 'They're *new* people.' We ignored the remark and another girl took over as spokesman.

'We normally play adventure stories of some sort,' she said. 'In the Infant School there was a painted frog on the playground.' 'It wasn't a frog it was a spider,' said another girl. 'It had eight legs.' 'Well anyway it was a pretend thing painted white on the playground, and it had little secret passages,' said the girl who was trying to explain; and the girl who had interrupted interrupted again to say, 'We used the middle of the spider for the igloo. We've been playing the Icy Game for a long time.' It takes me longer to understand explanations from 7-year-olds because they do not always use words in the accepted order. It is surprising how much difference there is between a 'painted frog on the playground' and a 'frog painted on the playground', even if the design in question was merely abstract and intended as a stimulus to imaginative play (in which it had clearly succeeded).

Rioting boys streamed round the corner. 'We're playing

Follow the Leader. We're playing Smack you on the Gob. We've got some more dirty jokes. There was this man went into a shop and said, "Have you got any dog's meat?" "Have you got a dog?" said the shop man. "If you haven't got a dog you can't have any dog's meat. You can't have any till you have shown me your dog." So the man went away and he came back next day and asked, "Have you any cat's meat?" and the shop man said, "You can't have any cat's meat till you've shown me your cat." The next day the man came back and he was holding a brown-paper bag. "What have you got in there?" said the shop man. "I've got a lump of shit," said the man, "can I have a couple of bog rolls?"' 'Where did you learn that?' I asked, in lieu of comment. 'I just thought it up,' he said.

'Here's a good one,' said another boy, and, in staccato sentences, like a gangster (partly because the bell had rung), he rattled off, 'There was this man and 'is wife died and 'e'd just buried 'er. And they said to 'im, "You left your wife's bum stickin' up." And 'e said, "Well, I got to park me bike *somewhere*."'

Yet another boy chipped in. 'Just time for one more before we scarper. You've got to write quickly. A man came into a shop and said to the shop man, "Have you got any women?" and the shop man said, "I don't think I have. I'll go and look in the attic." 'E came back and said, "Yes, I got one. It's 50p an hour." "I'll have two hours please," said the man, and 'e went up. He came down and said, "I'll be back tomorrow." Next day 'e said, "I'll have another two hours," and the shop man said, "All right, but don't squeeze the left tit." The third day the man came again and said, "I'll 'ave another two hours please," and the shop man said, "Remember not to squeeze the left tit." The man came down before his time was up. He was in a bad temper. He said, "I squeezed the left tit and I got my face full of rice pudding." The shop man said, "That wasn't rice pudding, that was maggots. She's been dead for the past twenty-five years."'

They tell the stories with a Rabelaisian gusto; the obscenities flow in a torrent. They are the unselfconscious conduits of an age-old, deep-flowing masculine tradition.

Wednesday 17 *October*

A silent day, blanketed in cloud. Later the cloud, which was really a vapour layer, cleared to reveal a tender blue sky—but that was after playtime.

The children were late coming out—no one knew why. I saw two eyes looking through the glass panel in the main school door, just topping the Cambridge-blue woodwork. By the time I opened the door the eyes had disappeared. I looked out of the door myself, at the same level, to find out what it felt like. From six handspans high the world looks much the same as it does from five feet three inches. The library is housed in the rather wide corridor (thanks to spacious Victorian planning). I read the list of library monitors, and looked at the Fiction section (mostly adult Penguin titles, such as Laurie Lee's *A Rose for Winter*). Then I noticed a few children wandering into the playground.

I shifted to the bench outside the door. Two dear little girls (the feminine principle personified) sat down beside me. One said, 'My Baby William is stuck up there on the roof. It's a sort of bean-bag doll, about six centimetres tall—yes, I would say about six centimetres. It comes in a matchbox. I'll show you.' She led me to the other side of the gable where Baby William, clad in a bright green romper suit, languished on the tiles. 'A boy threw him there,' she said. 'We must try to get him down.' She was placidly acquiescent, and gently amused. They began throwing small pebbles at Baby William. There was not the remotest chance of dislodging him in that way. (Children's first idea for dislodging an object from a high place is to throw stones at it; they do not seem to think in terms of climbing up or using long poles.) 'We must be careful the stones don't hit people when they come down,' said Baby William's owner, her round face heavy with responsibility.

Two girls were swinging like monkeys on the metal netball post. One of them said, 'Have you heard about the Irish punk rocker?' 'No, I don't think so.' 'He pulled the chain and 'is ear fell off.' I did not understand. 'Punk rockers wear a chain on one ear,' she explained. 'He pulled it like a lavatory chain.' She turned to a girl who was trying to pop a crisp bag simply by smacking it hard. 'Have some sense, blow into it first,' she said. 'I've done it like this before,' said the other defensively.

An unprepossessing 9-year-old with a pony-tail hair style, who has been lurking on the edge of the games-and-jokes-collecting crowds for several weeks, came and asked if I was writing everyone's names down. 'I'm not interested in people's names so much as the games they are playing,' I said. 'Well, I'll tell you mine. It's Lisa,' she said firmly, and offered me a crisp.

Two of the smallest girls were organizing sets of piggy-back players. 'Come on, get on,' commanded a horse. 'I will if you do it the comfortable way,' said the reluctant rider. The 'comfortable way' was 'with her hands under my bottom, not with her hands under my legs—that hurts'.

Suddenly I realized what was wrong with the playground. Nobody was playing conkers; yet on Monday, and before that, boys had been filling carrier bags with conkers from our great horse-chestnut tree. I confronted one of the boys: 'Why aren't there any conker games in the playground?' They were evasive. 'I got a bagful at home,' said one of them, looking everywhere but straight at me. Then he looked me full in the face and said, in the voice of one who has solved a problem, 'Want a joke?'

He was the boy I think of as 'the loblolly boy'. His arms and legs seem to move of their own accord; he flaps his eyelids over his large brown eyes when he pauses to think. He began: 'There was this man came and asked another man. He wanted to know where a sweetshop was and *'e* said, "Down the lane," and 'e said, "Where's this lane?" and 'e said, "In England," and 'e said, "Where's England?" and 'e said, "In the world," and 'e said, "Where's the world?" and 'e said, "In the galaxy," and 'e said, "Where's the galaxy?" and 'e said, "The Galaxy's in the sweetshop"—because Galaxy is a sort of chocolate, you know.' 'Lovely,' I said. 'Where did you learn it?' 'Oh, loads of people are saying it.'

'There was this Irishman, Englishman, and Scottish man,' said a girl who had been listening. 'Hang on,' I said. 'Yes, I'm ready.' 'Well, they were in a German camp and the Englishman was going to be shot by the Nazis and the Gestapo said, "What last word would you like to say before you get shot?" and the Englishman said, "Avalanche!" and the Nazis ran away and the Englishman was free. Then the Scottish man went to be killed by the Nazis and the Gestapo said, "What word would you like to say before you get shot?" and he said, "Blood!" and the Nazis ran away and 'e escaped. And then the Irishman went to be killed and they said, "What word would you like to say?" and the Irishman said, "Fire!" and they shot him.'[1] She was about 8. When she had finished she looked at me intently, her eyes widening with excitement, as if she was watching me undo a birthday present. 'That's a good one,' I said. '*Thank* you.'

A boy arrived with a problem. He was too tall and physically mature to have this particular problem, I would have thought, until I looked into his soft and childlike face. 'Hello,' he said. 'You know Mark Horrocks, the fat one? Well, *they* have gone off me, him, and a lot of others. I don't know why. *They* keep pretending I'm dead. When I talk to them they pretend they're hearing things.' He was really worried. 'They'll get over it,' I said. 'Try showing them they're screwy.' I screwed my forefinger into my temple. He smiled wanly.

The 8-year-olds in the queue provided me with a quick résumé of games I had failed to see during playtime. 'We bin playing Mother, May I? and Sly Fox. We played a little game of each.' 'Me and Karen's bin playing Chase.' 'We played Fairies and Witches.' 'You *didn't*, you need a rope for Fairies and Witches.' 'We *did*. We *did* have a rope.' 'Is that the game where one person twirls the rope round and round and everyone else has to jump over it?' I asked. It was.

Tuesday 6 November

A manic-depressive day; shifting black clouds, occasional shafts of silver light, a strong wind, and the pub sign creaking under the strain. Last week was half-term, when the Education Authority took the opportunity to paint some white squiggles, a

[1] When this story was heard in Manchester in 1991 an Englishman, Irishman, and Scotsman were about to be killed by bandits in a desert. The Irishman distracted the bandits by yelling his chosen word, 'Water,' and ran. The Englishman yelled, 'Hurricane,' and ran. It was the Scotsman who yelled, 'Fire,' and was shot.

spiral, and some bull's-eye circles on the playground, and some circles and cricket stumps on the end wall. When I arrived some girls were already running, mesmerized, along the line of the spiral figure, and at least a dozen smaller girls were jumping into the bull's-eye circles near the gate.

I was surrounded almost at once by the elders of the school—the girl elders, that is. 'Have you seen these?' they said, pointing to brooches they were wearing made from the metal caps and plastic inner caps of soft-drink bottles. 'You hold the metal cap against the outside of your jersey, and you push the plastic cap into it from inside your jersey. Coca-Cola tops are the best.'[1]

One of the girls rolled the collar of her roll-neck jersey up to cover her mouth. 'It's windy, isn't it?' I said. 'What did you do for Hallowe'en?' They had almost forgotten it. After a silence a rosy little blonde said, 'Someone came to our door and said, "Trick or Treat," and we gave them an Extra Strong Mint.' 'If you don't give them something they empty your dustbin,' said another girl succulently; and I suddenly understood how the threat of retaliation on the one hand, and the fear of it on the other, make the transaction vastly more dramatic and enjoyable than if it was simply a matter of begging and giving. 'And this neighbour of ours,' went on the first girl, 'she gave them 70p by mistake. She wanted to give them 30p, but she couldn't see in the dark and she gave them a 50p in mistake for a 10p.'

'What about Guy Fawkes Day?' I asked. 'Yesterday? oh, that was my birthday. I had these two to tea, and we had sparklers.' 'But on Saturday,' said her friend, 'she came out with us. We went to a fireworks display at a big house up the hill.'

Lisa had a friend in tow because she wanted to show me 'a new game'. She and the friend tangled themselves up. 'It's just called Tangling. There ought to be more people really.' Then they started dancing to 'a song we heard in the parish hall. It goes "Oom pa pa".' They were trying to be helpful.

[1] A correspondent told us she was making these in Manchester in the 1920s. 'I painted them and made them into badges for my Club, which met on the local off-licence doorstep, where there was a lamp. You just prised out the cork bit, put it under your jumper and the metal bit on top.'

We all looked at the smaller girls who were playing a game on the painted circles. 'What you do is, you jump from one circle to another. You try to jump into the circle in the middle and then you score fifteen. If you land in the outer ring, it's ten. If you land outside, you score five.' 'It doesn't seem right to score anything if you don't land inside the circle.' 'Well, you see, if *little* people are playing they can't jump very far, so they get five anyway. They add up the score as they go along.' 'So you think these painted lines are a good thing?' 'Oh yes. You make up your own games. You have to use your imagination—I heard someone saying that.' (But I understand there has been no directive from the headmaster, and no suggestions from the teachers.)

The loblolly boy brought forward a very small boy with cherub's curls. ''E's got a story,' he said. I had to bend down to hear it. 'There was this Irishman, Scottish man, and Englishman, and they all had dogs.' 'And *then*,' said the loblolly boy, scarcely able to restrain himself—'No, go on, *you* say.' The little boy went on, 'And the Scottish man said, "My dog can jump twenty feet high, so lick that," and the Irishman said, "My dog can jump over a double-decker bus, so lick that." Then the Englishman—then the Englishman said, "My dog's got a dirty bum, so lick that!"' The little boy looked up at me doubtfully. The loblolly boy put an arm round his shoulders and hugged him. They are not brothers.

The bell-ringer went round ringing the bell, and a girl with a raucous voice organized a last game on the spiral diagram. 'Nicola, Nicola, start in the mid-dul,' she shouted, and herself started from the outside. They met, scuffled past each other, finished the course and ran off to the classroom.

Thursday 15 *November*

'Raw', I think, would be the word for the weather this morning. The sky was pallid with cold. Poppy wreaths, with golden foil centres, glowed on the village cenotaph. Mars was being unwontedly thoughtful when he arranged for poppies to bloom in November.

Half a dozen boys came bouncing up with jokes at the ready. 'Mine first, mine first,' said the largest one. 'It says, "Doctor, doctor, I want a new bum." And 'e says, "Why, sir?" and 'e says, "'Cos mine's got a crack in it."' They collapsed with laughter and began whispering in a huddle. I thought an even ruder joke would ensue; but the next one was, 'What key won't open a door? A darkie.' Perhaps that *is* considered rude, nowadays. Then came the perennial favourite: 'What did the big chimney say to the little chimney? You're too young to smoke.' I tried to find out why it is so popular, but the teller simply looked into the distance and said, 'I got it from my brother.'

Lisa came up and said she had got another new game, and showed it to me with the help of a somewhat bewildered friend. There is something joyless about Lisa's 'new' games, which turn out to be combinations of parts of other games, or combinations of uninspired movements. She took the friend by the hand and pulled her so that they changed places; this was repeated; then they twizzled, holding hands; 'And then we fall down,' said Lisa—but they didn't. As choreography it was unimpressive; nor do I think it was invented for my benefit. As I walked away she was supporting herself on hands and feet, face to the sky, and telling the friend to sit on her stomach.

I stood watching the western side of the playground and thought, 'Playtime is a chance to go mad.' The children were pushing, clutching, staggering, prancing, dodging, exaggerating every movement into pantomime; it was continuous Saturnalia. One boy led another boy on a dog's lead big enough for an Alsatian. The tall bespectacled boy fired a giant elastic band into the air and tried to catch it.

A line of little girls, arms entwined, danced up to me and announced, 'We belong to the Abbey National. The whole *school* belongs. It's because the Abbey National man came to say, "Save Money." He gave us all these pamphlets and envelopes and we've got to bring 50p or 10p or 5p, and we put the envelopes in a kind of triangular box. Marilyn's not saving 'cos she's a grump.' Marilyn looked down at her toes and did not explain why she was not conforming.

The boys were whooping round us, riding on each other's shoulders. The mouse boy arrived saying, 'It's me, with Andrew on top. He's got a new way of getting down—look! Wow! I've trodden on his toes.' Andrew seemed a sober fellow, so I asked him about the outbreak of horse riding. 'We often do it,' he said. 'Why do you like it?' 'Sometimes I feel I'm falling,' he said. 'It's best to lean back.' (I have never yet had a real answer from a child to the question, 'Why do you like such-and-such a game?' Either they invent what they think is a suitable answer, or they shelve the question altogether.)

The children were screaming like swallows gathering in the autumn. The only quiet place was the eastern fringe, where the boys played their dedicated games of marbles. Play was especially serious today; the stakes were large chinas and rainbows.

At the end of playtime a little boy was discovered doubled up on the ground nursing a damaged hand. Two other little boys picked him up and put his arms round their shoulders, saying cheerfully, 'We're the ambulance men, we're the hospital men—come on then!'

Tuesday 20 *November*

A silvery winter's day, and as cold as steel. ('We must expect it at this time of year, mustn't we?') Three hours after sunrise an icy mist still held the village in subjugation.

The children were not in the playground at 10.35, so I went inside to keep warm. An ornamental young teacher stopped in mid-career from staff room to secretary's office and said, 'They're still in Assembly. It's really cold, isn't it? Perhaps they'll have some games to warm them up, this morning. Oh look, now they're out!'

Keeping warm did not seem to be uppermost in the children's minds. The boys flew to the marbles fence like flies to a flypaper; the girls stood round comfortably eating crisps and playing their mondial role of always being available for a chat.

One of them broke away to execute some dance steps in a circle. 'I'm trying to do the polka,' she said when she came back. 'We do that in ballet.' 'Oh, do you know *my* tune for it?'

I said, and hummed 'See Me Dance the Polka'. 'That's it,' she said, 'that's the one we use.' 'I used to dance the polka in a dancing class when I was younger than you,' I said. She took this as a challenge and said combatively, '*I* started when I was 2.' I had to give her best.

The air was filled with the sound of children playing. All around us machine guns stuttered, and the wounded gasped and groaned. Yet it was difficult to distinguish the fighters from other boys who were playing Stuck in the Mud, and I kept interviewing the wrong boy in my attempt to find out whether this was simply War rather than something more specific such as Germans and English.

'Why don't girls play War?' I asked the girls helplessly. They giggled at such a silly question. 'Because they don't *like* it. Because it's so boyish. Because they don't use guns,' they said vehemently, and with some disapproval.

'Well, how do you keep warm if you don't make war?' I said. 'We just run about. We go for a morning jog. We wear gloves—but we have to take them off to play ball games. They're playing Donkey round the corner, did you know?' We went round to have a look, and then settled cosily on a bench. Gloves, like the famous conversational gambit 'Do you like string?', proved a rewarding topic. 'You can throw your glove up and try to make it stick on the wall,' said the 9-year-old who looks like a dark-haired beauty on an Edwardian chocolate box. 'And there's that game with chocolate that people play at parties. You have to put gloves on and cut up the chocolate with a knife and fork.'

They scanned the playground, and tried to console me for the lack of interesting games. 'There's always football,' they said. 'And the boys are playing that game of throwing a ball over the roof and catching it the other side.'

All the time Lisa had been tapping me on the arm and trying to show me a routine she had invented with a miniature bouncing ball. 'Drop it, and catch it, and then wave your finger like this,' she said. 'Get it?' I longed to tell her she is a bore. She seems to have been born without any idea of what is interesting or not.

The loblolly boy took me over to an angle of the wall, where a small boy was crumpled up, crying. 'Write that down,' he said. 'He's crying,' I said, 'that's not a game.' 'Fighting's a game. He's been fighting.'

Wednesday 28 *November*

Officially 'mild'; but the grey mildness was laced with a cold wind.

An urchin tore across the playground, tossing playing cards to right and left. 'They're car cards,' said the boy who was hastily picking them up. ''E wants to get rid of them and 'e knows I want them.'

'The speed of modern living' is a cliché of unproven truth; but the speed of children's living has never been in doubt. A fifteen-minute playtime can be packed with at least twenty brief and random incidents.

A boy stopped briefly to ask, 'D'you like my frog? He's full of beans—actually I think it's split peas—I'm almost sure. You can feel them inside him.' He raced on to show the frog to another friend, and instantly a girl was telling me, 'It's my birthday, and look at my bear I've been given by my parents. He's called Sucking Bear on the box because he sucks everything—his toes and his thumb—you push them in his mouth, see? I haven't given him his real name yet.'

'Miss, you watch us,' said Lisa's bullying voice. 'Here's a new game.' Lisa took her long-suffering crony by the hands and, heads wobbling, chanting, 'Punchie girr, punchie girr,' they danced erratically into the distance.

The big craze surrounded me suddenly like a whirlwind. Boys were striking attitudes, holding little red and white objects between forefingers and thumbs and letting them fly. 'What are they?' I cried enthusiastically (because they were nice bright little objects, and new). '*I* don't know,' said the genial ginger-head (a future lorry driver?). 'You get them in the stationers. They cost 30p for ten. You pinch 'em and let 'em jump.' 'They're called Classroom Pops,' said the equally genial tall boy with spectacles. 'They *could* be called Classroom Pops, anyway.' I said, 'They look like those tiny Victorian cotton-reels—red ones, with white cotton.' 'They're foam plastic,' said the boys severely. (I bought a packet afterwards. 'Squash Pops', they are called: 'The New "Squash and Pop" Foam Missile. Hours of Fun for Everyone. Instructions: Hold the Squash Pop between forefinger and thumb and squeeze until the Squash Pop "pops" from your fingers. See how you can improve your aim with practice. Keep out of reach of very young children.' Made in Scotland by Thomas Salter Toys, Glenrothes, Fife, copyright 1978, who suggest 'just two games to play', Tiddlypops and Target Pops.)

The potential lorry driver remembered he had a story for me. He began telling it, rhythmically:

> There was a man
> 'E went in a pub
> 'E said to the man behind the bar,
> 'Can I have twelve pints of beer, please?'
> And 'e drank them all.
> Then 'e got in 'is car and 'e was driving home.
> And the police stopped 'im.
> 'Excuse me, sir, you was doing a hundred and ten miles an hour, wasn't you?'
> 'Yes.'
> 'Why was you doing a hundred and ten miles an hour?' [in a deep solemn voice]
> 'So I could get home quicker before I caused a crash.'

His Dad told him that one, he said.

The whole group came with me round the playground. We saw a few marbles players; and someone lying dead; and some girls playing Mother, May I? in the alley-way; and a great number of boys playing football with large glass marbles—'because there's no football, because someone smashed a window—we'll tell you who it was, if you like'.

Three girls stayed with me, and a boy wearing a false nose and spectacles. 'We'll tell you some more games,' they said. 'We play Its on the netball pitch. Someone's on it and there's lots of people and they're not allowed out of the square. You play it on a quarter of the netball pitch, and there's *no* home and *no* Scribs [truce term].' 'There's Red Rover—I'll tell you how to play that. There's a line holding hands on one side and another line the same on the other side and one side calls out, "Red Rover, Red Rover, call *somebody* over," and that person has to try and bash through the other line, and if they get through they bring someone back for their team, and if they don't bash through the line they stay on their side.'

'Knock, knock,' said one of the girls suddenly. 'Who's there?' I answered. 'Egbert.' 'Egbert who?' 'Egg but no bacon.' A girl with wide serious eyes said carefully: 'What has two heads, six legs, four eyes, and goes to football matches?' 'A mounted policeman,' said the boy swiftly. Then a jolly little girl butted in with a string of 'Knock, knocks': 'Arthur . . . Arthur any biscuits?'; 'Loraine . . . Loraine falls out of the sky;' Celia . . . Celia lips and don't tell a soul'. 'Knock, knock. Who's there? Irish stew. I arrest you in the name of the law.'

The bell had gone and more boys were clustering round with jokes to tell: 'Where does a fish keep its money? In a river bank;' 'What do you call a camel at the North Pole? Lost;' 'What's purple and burnt the cakes? Alfred the Grape;' 'What's the smallest book in the world? *The Irish Book of Knowledge*;' 'What animal can jump higher than a rocket? Any animal—rockets don't jump, they fly;' 'Did you hear about the cowboy who wore paper clothes? He got arrested for rustling.' There is no difficulty about collecting jokes; all one has to do is to stand still and try not to be swept away in the flood.

At the end of this spate a small girl rattled off a joke so quickly I could not hear it. A motherly bystander said primly 'It's rude. She shouldn't say it, really.' The others were still dancing about and chanting, so I bent down to let the child say the rude words to me privately:

> 'Knock, knock.' 'Who's there?'
> 'Twelve.' 'Twelve who?'
> 'Twelve and twelve are twenty-four,
> Shut your mouth and say no more.'

I was spared from comment by the arrival of the teacher. Suddenly the children were sober and standing in straight lines. 'Fourth-years!' said the teacher from the top of the classroom steps. 'Will you please quietly and efficiently get your maths and go to your maths sets?'

Wednesday 5 December

A wind strong enough to lean upon; and damp, too.

The teacher seeing the 7-year-olds across the road (wearing a pale green mackintosh and mulberry-coloured stockings) was using a broomstick as a staff. 'They need it for the Nativity Play,' she said. 'I don't know why.'

I wrote slowly in my notebook (a disengagement respected by the children) and listened to the savage noises of the playground: the aggressive and defensive shouting—'Nah, gettaway,' 'Watch it, you dope'—and the wordless invective of screams and snarls. Hobbes, I thought, was right in his belief that man is committed to endless conflict.

Two boys had been waiting patiently to deliver a joke. They often stand in pairs, waiting; one to tell the joke, and the other as sponsor. When I looked up the taller boy said, 'Doctor, doctor, I keep thinking there are two of me. One at a time, please,' and marched straight past without waiting for my reaction.

Lisa is wearing her hair loose round her face, and has a fringe. She has been transformed into quite an attractive child. But she is just as possessive, in fact a damned nuisance. It is 'Miss, watch this!' all the time. She showed me several boring sequences with a ball today, just throwing it up and catching it and passing it round her body. She invents these things especially for me, so that I will write them down. I am sure she would not play them otherwise.

I was rescued by a girl who has dark, mournful eyes in a narrow face and who, as far as I know, never smiles. She seems to prefer giving me a running commentary on the games other people are playing, rather than joining in herself. We began strolling round the playground. ''Ello,' said the cheerful curly-head. 'We're playing Sly Fox'. 'How you play Sly Fox,' said my escort, 'is—he stands by the wall and keeps turning round and if he sees them moving he has them, and they stand by the wall, and the next one had has to hold the first one's hand till they're in a chain. And when they're all had, he tells a story—it can be about anything you like—and as soon as he says, "Sly Fox," he chases all of them and the one he has is on it.'

Sly Fox was all over the playground: great big games, stretching from one side of the football area to the other. 'It's because they're still not allowed to play football,' said the girl. 'And over there they're playing Chasing, and over there they're playing piggy-back fighting.' 'Chiggy-Back is a better name, *I* think,' said a girl who had overheard. She herself had been playing the casual, lifting game played by generations of dilettanti in this playground: 'You put your hands together in a hand-grip, and the other person does the same, and you run and sort of hook on to her, and she lifts you round.'

The lorry-driver boy and his gang swirled around me, laughing stridently. They were mobbing the boy with the crew cut. 'He's funny,' they choked, pointing at him. 'Is it his hair?' I asked. 'No, he's just funny. He loves Julie Davies. And he laughs funny.' The victim, though pink in the face, seemed as embarrassed for his tormentors as for himself. We exchanged adult glances, and shrugged our shoulders.

The lorry driver suddenly stopped and became serious. 'I got a joke,' he said. 'There's this Irishman and 'e goes into a bakery—' 'Where *do* you find these stories?' I said, all femininity. He looked pleased. 'I find them in my pocket,' he said, and looked round for applause. 'Well, this Irishman goes in the bakery and 'e says, "Can I have a brown loaf, please?" and the baker says, "Sorry son, we only got a white loaf" and "Oh," he says, "that's all right 'cos I've got my bike outside."' 'It's good, en' it?' exulted the others, reeling under the excellence of the joke. When they had recovered their balance and once more stood before me, another boy took up the story-telling. 'Do you know about this Irish footballer?' he yelled. 'This Irish footballer, 'e went to this man who asked questions, and 'e goes, "What's your occupation?" and 'e goes, "Engineer," and 'e says, "What's your hobby?" and 'e says, "Footballer," and 'e says, "What do you do in football?" and 'e says, "Pass."'

Having delivered his punch line, he grinned like a ventriloquist's dummy, baring his teeth and switching his head from side to side. The story was in imitation of the television quiz *Mastermind*.)

A more sedate lad, not belonging to the fraternity, began telling a story. 'There was this Englishman, Irishman, and Scotchman wanted to go to the Olympic Games, and they hadn't got tickets, and they tried to force their way in.' (The gang encircled him, ruffling his hair and jabbing him in the ribs as he laboured.) 'The Englishman put on a tracksuit and got a javelin and said, "I'm English and I'm in the javelin competition," and they let him in. And the Irishman, 'e got a tracksuit and 'e got a barbed wire fence, and 'e said, "I'm Irish and I'm going in for the fencing competition—"' I never heard the end of the story because the teller was so overcome by its funniness that he threw himself backwards on the ground with his heels in the air.

Suddenly they realized that the other children had been lining up for some time to go into class. 'Oh gawd,' said the lorry driver, and ran helter-skelter for the lines. 'If you're looking for trouble—' he murmurered as he ran, and when he had recovered his breath he recited:

> If you're looking for trouble
> You've come to the right place;
> If you're looking for a fight,
> I'll smack you in the face.

'It's an Elvis Presley song,' he said as he disappeared into school, 'but 'e didn't put in that last bit—*I* did.'

Wednesday 12 *December*

The sun, like the villagers, was reluctant to come out. (In midwinter the village does not become really crowded until about 11.30, when the queues lengthen in the butcher's and greengrocers' shops and remain long until nearly one o'clock.)

I still feel the excitement of the hunter when I hear children coming out to play; and the children—especially the boys—

seem to hear a 'tally ho!' in the air as they come running into the playground. It is a general response to life, in their case, as they launch themselves into fifteen minutes seemingly coloured by feelings of 'What's new?', 'Anything might happen!', 'Up me, down with the other chap!'

The boys scrimmaged cheerfully by the gate, using the time to garrotte and cuff each other until some more definite activity presented itself. The girls were caught up in the flurry. ''E hit me, sausage roll face!' one of them called out delightedly, and chased the boy in question. Another girl said, comfortably, 'We're having a nice Christmas party on Friday.'

A boy brought a plastic-paper fish in a tiny envelope. 'This is a fortune teller,' he said. 'Hold out your hand and I'll put it on your palm. It says if the tail and the head go up towards each other it means you are in love.' I was very much in love. 'We had a Christmas tea yesterday,' he said. 'The fish was in my cracker. We each have a Christmas party, each year of us. It was the second-years' yesterday.' A little girl looked up at me and said 'What does a plumber need? A leak.' I wondered if it could be a cracker joke, and decided not.

The school's most ardent practical scientist was hurling a huge feather into the air time and time again. 'We're testing it,' he said. 'It's a pheasant cock's feather, and we are testing it to see if it flies. So far it done two loops the loop—and it fell down there.' Typically the feather did not belong to him but to a quiet boy who said his father had found it in a field while taking the dog for a walk. The scientist went on enthusiastically, 'There's another thing it does. You pull it to pieces like that, and it does up again like a zip.' One of nature's best gadgets, a feather.

I was adopted by a boy with shining eyes, who plied me with jokes and stood on tiptoe to watch me write them down. 'What is a teacher's best fruit?' he asked. 'A date—in history, you see. What was the egg doing in the jungle? He was egg-sploring.' 'Hurry up,' said a girl to him, irritably, and went away. He continued, 'Here's a funny one. There was this boy who said, "Mum, I think my teacher's in love with me." "Why's that, son?" "Because she keeps putting crosses by my sums." I got that out of *The Target Book of Jokes*.' 'Why was that girl so impatient for you to come?' I asked him. ''Cos she wanted me to play Kiss Chase. They're all doing it this morning. And those lot over there on the ground, they're playing Love War.'

He went off a little reluctantly to fill his social role, and his place was taken by two senior girls. One had the other firmly in tow. 'Could you write down Kathleen Marsh for Craig Newman and Darin Phillips, please?' she said, while Kathleen looked suitably sheepish. They then became more adult, or was it less? 'Have you seen *Mind Your Language* on the television? You should watch it, it's good.'

All this while I had been aware of Lisa and her persistent 'Look Miss, look Miss!'—but I gave no sign of having heard.

The boys, at a loose end because football has been stopped, were either playing marbles, or fighting, or being chased and fought by the girls. A group of boys, including the 'lorry driver', came swarming round wanting to tell jokes; but beyond them I could see two girls hand-clapping with lackadaisical, minimal hand movements and bored faces. They were scarcely attending to the game, and the big golden-haired girl with the sulky mooncalf eyes was looking into the distance with a dissatisfied expression. 'What's the rhyme you're using?' I asked brightly. The smaller girl was reticent. She said slowly, 'It's a funny rhyme, really. It goes,

> See, see, my baby,
> I cannot play with you,
> Because I've got the flu,
> Chicken-pox and measles too:
> Flush down the lavatory,
> Into the drainpipe,
> And that's the way they go—go—go.'[1]

'*We* didn't make it up,' she said defensively. 'It's going round the school.'

The boys had been throwing themselves at the girls to break up our conversation, and now lined up in an orderly manner to tell their jokes. 'I got a good one. There was this boy and 'e goes, "Mum, Mum, what's a werewolf?" and she says, "Shut up and comb your face."' (I had not heard any 'Oh Daddy' or 'Oh

[1] See I. and P. Opie, *The Singing Game*, 474–5.

Mummy' jokes since the late 1960s, and then they were told mostly by students and office workers.) Even quite a neat 'Knock, knock' seemed feeble by comparison, giving verbal titillation but no breathlessly wild image. 'Knock, knock.' 'Who's there?' 'Senior.' 'Senior who?' 'Seein' yer so nosey I won't tell you.'

Miss —— came striding through the crowd shouting, '*Get* in *line*,' and I crept away. I finished writing my notes in the company of a left-over marbles player who was intent on getting his marble into the hole.

Tuesday 15 *January* 1980

The winter term began yesterday. The playground was covered in melting frost, and a thin fog hung over the village.

The children were in full scream by the time I arrived. I went inside and leant against the gate watching them. A boy was already propped up there, laughing at the scallywags trying to play Budge He on the painted circles and unable to keep their balance. 'It's a mixture of chasing and sliding,' he said. 'Look at him, he keeps falling over 'cos he's got ice on his shoes.'

In the distance I saw the editor of *The Form Three Times* holding his arms out sideways. (*The Form Three Times* is back in production after nearly a year. I am to be interviewed for it soon, I understand.) 'We're playing Stuck in the Mud, but we call it Stuck in the Glue—Stuck in the Glooo,' he said, savouring the extreme stickiness of the word. As he was stuck, he was free to talk. 'You play it by having somebody on it, and they chase the others and when you're had you put your arms out and somebody has to free you. Gavin and Derek thought of it. We were playing with them at the time anyway, so we had enough people.'

Other boys congregated, not part of the game. I said, 'I've been reading about some children in Oxford who bang their foreheads with the palms of their hands and say, "Taxi," when they've made a fart.[1] Do you do that?' (They were not surprised

[1] I had been reading Andrew M. Sluckin, 'Avoiding Violence in the Playground', *Educational Research*, 21. 2 (Feb. 1979).

at what the Oxford children do, but I think they were surprised at my talking about farting.) 'No, no,' said a tousled urchin who is almost a clone of Richmal Crompton's William. 'We say,

> Beans, beans, make you fart;
> Beans, beans, are good for your heart.
> Apple crumble makes you rumble;
> Custard powder makes it louder.

And loads of people, when somebody's done a fart, they keep going like this [jabbing down with a thumb] and going [i.e. saying], "Plug! plug!" and the last one to do it—they have to eat it.'

'Can I say one now? It's rather rude,' said an urgent voice at my elbow. 'It's about two men who were working down a manhole, and one of the blokes goes, "I'd better fill your flask up 'cos you're working down here all day," and the other man says, "Don't drop the flask 'cos you'll have to buy me a new one," and the man does drop it and the other man says, "Go and buy me a new one, it's called Duraflex and it's about that long." And so he goes, "Right," and 'e goes to the shop and says to himself, "What was the name of what I want? Dura—ex—Durex, that's it!" and 'e says, "Can I have a Durex, about this length?"' Here the boy broke off, looked at me sideways, and said, 'Do you know what a Durex is?' 'Well yes,' I said, 'it's a condom;' and then, relenting, 'It's a French letter.' He looked at me scathingly. 'It's not—it's a rubber Johnny.' He finished his story. 'And the shop people were surprised. They said, "What do you want a Durex that length for?" and the man said, "There's two of us working down the same hole."' The audience cheered. They were all 8 or 9 years old. I think they get this kind of story from their fathers, who get them from the ubiquitous and monothematic joke books on sale in newsagents.

I thanked the storyteller warmly, and the boys went happily off to queue outside their classrooms. Some girls whirled past me, highly excited, giggling, 'Me and Karen's been smacking the boys' bottoms.' We need not wait for spring or maturity, it seems, for youthful fancy to turn to thoughts of love.

Tuesday 22 January

The sky was a wide expanse of blue (bluer than usual in January) with white cloud-drawings across it, and a long line made by an aeroplane exhaust. Windy, and quite cold.

I was already leaning against the sunny south-facing wall when the boys erupted from their classrooms. One of them stood consulting his watch two feet away from me, then said crisply, 'John, run round the block. Start from now, right?' When John had finished the circuit I asked him, 'Are you a good runner?' 'Yes,' he said, 'when my brother is going his fastest on his bike I can win him.' He went off again, wobbly as a new-born giraffe and weaving from side to side.

There was an Olympic tone in the playground this morning. Training shoes and track suits (with Snoopy Dog badges) were much in evidence. I thought of raising the subject of the Moscow games, but decided it was outside my province. One of the footballers came and said, 'Get out of the goal, please.' The runner and his trainer and I moved to the corner of the wall. 'Do you play football?' I asked the runner. 'I don't play here, and I can't play at home, either. My Dad is a footballer. He was the Welsh goalie. I'm going to be 8 this year—in February.' ('When I'm 8 I'm going to ride a motor bike,' said the friend. 'He's not, he wouldn't be allowed,' said the runner.) My mind busied itself among these statements, seeking a conclusion. 'Does that mean that when you are 8 you will be able to play football?' I asked. 'Is that the rule?' 'I don't know,' said the runner, 'but it is usually the fourth- and third-years who do it.'

My relationship with the pestiferous Lisa has worn thin. This morning I found myself refusing a potato crisp and the offer of one of her creative game demonstrations.

The ginger-haired lorry driver planted himself in front of me. 'Miss, how do you tell the difference between a snowman and a snowwoman?' he said. 'I don't know.' (It is quicker and safer not to know.) 'Snowballs,' he said, briefly. He got it from his Dad, he said. Now we were well away. 'How d'you get paper babies? From an old bag. Did you hear about Lieutenant Kojak?

He went swimming and his behind was sticking up and another man said, 'Look at Lieutenant Kojak, he's split his head." '[1]

All the time, little boys had been creeping and peering round the corner, almost under my elbow. 'I expect they are playing Om Pom or I Spy,' said the lorry driver paternally. 'Wot you playing?' he shouted, but they were too engrossed to reply.

'I got another joke,' said the lorry driver. 'There was this Irishman goes to the baker and 'e asks for a brown loaf—' 'I'm sure I've had that one before,' I said desperately. 'Yeh, I probably told it to you,' he said, going on with the story. I looked into the distance, checking as well as I could on the games: two groups of marbles players, no skipping, no bottletops.

'I got another joke,' said the lorry driver (perhaps he will not be a lorry driver, perhaps he will drive a bulldozer). 'I'm supposed to be writing down games, really,' I protested faintly. 'My Dad got some good jokes,' said the 7-year-old runner. 'You can get joke books, and that,' said a girl. 'I always make up my jokes myself,' said the runner. 'This joke is *about* games,' said the lorry driver, cutting in decisively. 'There was this man and 'e went to a fête and 'e has a go at hitting playing cards with a dart and 'e wins so 'e gets a tortoise. And 'e comes back a year later and has another go. And 'e wins and the man on the stall says, "What would you like this time?" and 'e says, "Anything—but I wouldn't mind one of those crusty pies again." ' The girls crooned their appreciation.

'I *must* get some games,' I said, and walked resolutely towards the only structured activity I could see, near the dustbins in the corner. A tall girl with a mop of bright red curls like a 1920s boudoir doll was standing holding a long piece of knicker elastic. 'We're playing Dogs,' she said. 'Me and Clare are the dogs and Vanessa and Chris are the owners.' 'What kind of dogs are they?' I asked the very small male owner. 'Any old dogs,' he declared stoutly. 'We give them chocolate biscuits and chocolate buttons.'

I went to the alley-way where the children were lining up for the outlying classrooms. 'What about games?' I asked the girls.

[1] Lieutenant Kojak is the hero of a popular American detective TV series. He has an entirely bald head.

'Can I tell you a joke I made up last night?' asked a little lost boy who had been following me around. He looked up at me with huge trusting eyes and said, 'How do you know that a car has got cancer?' 'I don't know,' I said. 'When it is smoking out of its chimney,' he said, and went away satisfied.

'So—what games have you seen? I haven't got many written down,' I asked the girls again and was again interrupted. This time I was slain by a Sound Phaser, a magnificent plastic gun which showed a pinpoint of red light and squeaked when the trigger was pulled. 'It doesn't squeak,' the boys protested. 'It's got two sounds, a Sonic Destruct and a Laser Blast.' 'It's terrific,' I said, and meant it.

The girls said they had seen people playing Semicircles. 'It is played on the semicircles of the netball pitch, where you shoot from. They use them for a home, and one person is on it and the people have to run from one side to the other without being caught.'

All through the explanation I was being assaulted by a six-inch dirty-white football. A small boy was presumably over-excited and, like the Russians in Afghanistan, trying to find out how far he could go without retaliation. I came away with my cheek smarting, wounded on active service.

Wednesday 30 *January*

I was in a rush and could not even think how to describe the weather; so I asked the 7-year-olds, lining up in their playground. 'It's windy, cloudy, and ice-cold,' said a coffee-coloured curly-head. Their teacher came to take them over the road; 'All right. Quiet now!', and we trooped over. One of the boys slipped out of line to tease the boy who was leading. The teacher called out, 'Come back, David, come and hold my hand—you're being silly.' 'Ooh, it's not fair,' whined David from the far pavement.

It was not ice-cold. The light, damp breeze was as refreshing as a touch of eau-de-Cologne on the nape of the neck. I was further revived by a snatch of song I had not heard for at least five years:

> There was a man in '92, parlez-vous,
> There was a man in '92, parlez-vous;
> There was a man in '92,
> Did a fart and caught the 'flu,
> Inky pinky parlez-vous.

The boys singing the song were alight with glee. They did not know what 'parlez vous' means. 'It's only a load of old rubbish, but we like singing it,' they said cheerfully, and went rollicking off with their arms round each other's shoulders. 'Mademoiselle from Armentières' has a good enough tune to carry parodies even poorer than this.

The William boy and a friend came skidding round the corner, wanting to tell jokes. 'I can prove to you that I have got eleven fingers,' said the William boy and spread his fingers out. 'You count ten, nine, eight, seven, six—No, shut up [to his friend, who was trying to join in], it's my joke. You start on this hand, ten, nine, eight, seven, six, and then say, "Plus five is eleven."' They both began shouting at once. 'There was this Englishman, Irishman, and Scottish man and they went into a haunted house and there was only one more room left and they suddenly heard a voice saying, "Got you, got you! Now I'm going to eat you," and they went in and there was a boy sitting on a chair with his finger up like this—it was a bogy, see, from his nose.'

A scurry of girls arrived. 'Will you write, "Paul Colston is a fool," and "Richard Williams is a twit"? Go on, *please*. Put it in the paper. They've been laughing at our song we've been making up. We learnt it from Tracey. She had it from her big sister. We knew a lot of it already, but we learnt the other little bits. We learnt it last week, and we kept doing it and then we learnt other people. We'll show you.' They stood in two lines facing each other and went into a song-and-dance routine:

> In and out of the red balloon [move toward partner, clap hands together, retreat, describe balloon in air with hands]
> Marry the farmer's daughter [link arms with partner, go round and back to place]
> Sleepyhead in the afternoon [hands together as pillow, rock head]
> Britannia, Britannia [fling right arm round and back, then left]
> Melissa [both arms round and back]

Sleepyhead in the afternoon [as before, then turn facing outwards]
My—name—is
Diana Dors, I'm a movie star [jump round to face partner again]
I wear long johns and I play the guitar [suitable motions while swaying to rhythm]
I've got the legs, the legs [legs forward in turn]
The hips, the hips [smooth the hips each side]
Turn around movie star, hey hey! [twirl round, throw hands in air at 'hey hey!']
See those girls in red, white, and blue [move forward, winding fists round each other]
See the boys say 'How do you do?' [return to line, unwinding fists]
So—keep the sunny side, keep the sunny side, keep the sunny side up, up,
And the other side too, too [jump, crossing and uncrossing the feet, kick legs in air at 'up, up' and 'too, too', clapping under legs]
See the soldiers marching along,
Elvis Presley singing a song,
So—keep the sunny side, keep the sunny side, keep the sunny side up, up,
And the other side too, too. [actions as before]

'The words mean nothing, really. We just made them up.' But this applies only to the first six lines. 'My name is Diana Dors' has been a popular dance routine since at least 1966, and 'Keep the sunny side up' since at least 1960.[1] It took some time to set down the words and actions. They could not think of either without performing the dance from the beginning, and I could not write and describe the actions for more than two lines at once.

Friday 8 *February*

Moist and mild.

A pygmy figure rambled into view, hitching up his long trousers. (The smaller boys are perpetually in difficulties with their long trousers; it gives them a vulnerable, Chaplinesque

[1] See I. and P. Opie, *The Singing Game*, 415, 429–30.

look.) More boys came out, squealing, eager, earnest; in search of fun, or trouble. They formed a semicircle round me, not sure who would start the riddling session.

A fair-haired boy took the lead (a truly British lad, with wide blue eyes). 'What's red and dangerous?' he asked, and answered himself, 'An avalanche of apples.' The riddles streamed by, complete with answers. But at least I used to know the answer to 'How do you make a Maltese Cross?' ('Strike a match and stick it up his jumper') and I was trying to remember it when I was struck in the back by an elastic band. 'Do you know what my hobby is?' asked the William boy from behind me. 'Firing crossbows at targets—mostly it's at the girls.' He had a supply of elastic bands in a red and yellow Lego box. 'Oh 'eck, you're not going to write down that I did that, are you? Bloomin' 'eck!' The whole crowd broke into a chorus of 'A-a-a-a,' clenching their fists and pummelling each other. 'We're tanks,' they said.

A fourth-year girl came to fetch me as a witness. 'Will you write down, "Linda Harris is always kicking?" She kicks like this, forwards, and she kicks backwards as well.' 'It sounds as if you have been playing the Okey Kokey,' I said to Linda. 'We have,' she said plaintively; and then added with enthusiasm, 'And we've been playing Orange Balls—you know, you go in a ring and they sing, "Orange balls, orange balls, here we go again," and the last one to sit down, she goes in the middle. Then they pick her a sweetheart and she has to clap her hands if she likes him—but usually she stamps her feet because she hates him. I like that, it's fun.'[1]

The girls gathered round, wanting to help with some games-reporting. 'We've been playing Semicircles on the netball court, and we've been playing Blobs, you know, in those circles on the playground. We jump from circle to circle and someone has to try and catch you when you're not in a circle, and you go really fast like Keep the Kettle Boiling, and sometimes if someone's on a circle you push them off.' 'They are called Blobs, are they?' I said. 'Well, *we* call them Blobs,' said the girl, with the accuracy I so much appreciate in children.

[1] Ibid. 232–5, 491–8.

A tremendously loud chanting arose in the corner of the football area, where the boys were leaving their game. After consulting at least ten of them ('I don't know it really,' 'I don't understand it at *all*,' 'He knows it, I *don't*') I wrote the sounds down as,

Oggie, oggie, oggie,
Oy! oy! oy!

'I think it's Welsh,' said one boy. 'I think it's the Welsh way of saying "Liverpool are magic."' More research seems to be needed. They went away in their usual euphoric mood, fitting in one more jollification before the discipline of school closed in upon them. 'We don't need no education,' they sang, 'Hey teacher, leave the kids alone. All in all, Was just another brick in the wall, One, two, three—scream!' 'That's a pop song,' they said. 'It's in the Top Twenty.'[1]

The William boy was late coming in from play. He ran to the classroom steps and shot a last elastic band in the air. 'Shuttle goes up, and space ship comes down,' he said. He glanced in my direction. 'See you, then,' he said, and disappeared into school.

Tuesday 12 *February*

Nippy. Spears of sunshine pierced the clouds.

I read the school notice board while I was waiting. The Chamber of Commerce were advertising:

Pancake Races
Saturday 16th February
2.30 p.m.

*Cash and other Prizes Galore
*All Children Completing Race Receive a Prize
*Children and Adults Races

Not a word about Shrove Tuesday. Nowadays the emphasis is on pancakes (the BBC calls the day 'Pancake Tuesday').

The birds had been shouting all the way down Hunter's Hill,

[1] Pink Floyd, 'Another Brick in the Wall'.

and I went to the marbles fence to wait for spring. I could see a long string of girls attached to the school wall, playing Sly Fox, and a machine gunner lifting his head from the ground to say 'Ah-a-a-a-a'. A boy came close to me, hesitated, murmured, 'No, it's not a very good one,' and went away again.

Two cherry-cheeked, breathless lads crashed into the fence and told me 'We're playing Fish and Chips.' They said, 'One person is on it, and they try and get the other person and you try to get away before they can eat you all up. I don't know why it's called Fish and Chips, especially. We just call it that. I like eating Fish and Chips.' 'Nobody's playing marbles,' I said. 'We don't play it in the winter,' he said, and rushed off to tease the On It—'Daisy, Daisy, sitting on the wall.'

'What was that he was shouting?' I asked the mouse boy, who had been waiting quietly with a friend. (I am perpetually in trouble, not knowing what is going on, picking up fag-ends of passing rhymes, having to be helped.) 'Don't know,' said the mouse boy. 'People call the chaser names until he chases them, like "Who's a little fairy cake then?"' He wanted to talk about his own, more interesting, business. 'We've got a new weapon. It's called a "Q Bull Head". You make it like this. There's a weight—made of Blu-Tack[1]—and it's on the end of a sharp piece of plastic. When you throw it it goes right into someone's arm.' 'Is it a knitting needle?' 'No, my Dad made it from a piece of plastic. It's called Q because we've got this Q Club. We do karate and fighting stuff and that. And there's a sort of Spy Club. Us two made up the club together. It's got a lot of clubs in it, so you've got a lot of clubs all together in the one club. There's only six people in the Q Club. We have an election for the leader. But we go round asking *everybody* who they will vote for out of the six.' The mouse boy has become very earnest and responsible since his friend moved to another school.

Two delicious little maidens ('little blossoms' they would be called in Yorkshire) had been hand-clapping within earshot while we were talking. I had heard 'See, see, my baby', and then:

[1] Bostik Blu-Tack is a re-usable, putty-like adhesive.

I'm Popeye the sailor man, full stop,
I live in a caravan, full stop,
And when I go swimmin'
I kiss all the women,
I'm Popeye the sailor man, full stop,
Comma comma, full stop.

I asked them if I had got the words right, and other girls crowded round. An older girl said, 'There's more words than that. There's—

I'm Popeye the sailor man,
I live in a pot of jam,
And when it gets sticky
It sticks to my dicky,
I'm Popeye the sailor man,
Comma comma, dash dash, full stop.'

A thin-faced girl with wide navy-blue eyes was delighted to contribute another version. She had wanted to be sociable, earlier, but could not think of any conversation. She recited,

I'm Popeye the sailor man,
I live in a caravan,
I live with my granny,
She showed me her fanny—
I'm Popeye the sailor man.

'There's another rather rude one, too,' she said, and sang, to the old tune 'Country Gardens',

What do you do when you can't find a loo,
In an English country garden?
Pull down your pants, and suffocate the ants,
In an English country garden.

She pointed at a tall blonde friend and said, 'Her brother was saying it to Julie, and I kept on being *amused* with it and I spread it round the school.'

The William boy had been agitating at my elbow. 'Can I tell a joke, can I tell a joke? Oh blimey, oh quick the teacher's coming!' 'When is a bus not a bus?' he said. 'When it turns into a street. And there's another one. Why was the policeman up the tree? He was in the Special Branch.' He never waits for an answer and, having delivered himself of his burden, ran off happily to his classroom.

The tiny maidens had been waiting patiently. They had a rhyme to say. The most cherubic of the two dictated it slowly (with her rosy cheeks and fair curls she looked as if she had stepped out of *Father Tuck's Nursery Favourites*), and looked over my notebook as I wrote it down:

Aaron and Moses were down the duck hole,
Aaron set light to Moses' bum hole;
Moses jumped up in a hell of a fright—
'Aaron, you sod, my bum is alight.'

'My Dad taught it me,' she said. I wondered what kind of a Dad she had. 'He works on a farm,' she said. 'He lives on one. He drives a tractor, and he milks the cows.'

Tuesday 26 February

A sharp frost still melting on the grass; and the sun shining through the fog like an outsize silver penny. The gang of workmen on the railway showed up as shadow puppets, and the villagers, though not actually groping, seemed bemused. At the corner by the school, council workmen in orange waistcoats had removed a manhole cover and were stirring evil black sludge with iron staves, laughing as they did so.

The playground seemed less foggy, perhaps because everybody was nearer the ground. A little girl with red-gold hair (a miniature version of Rossetti's Beatrice) was trying to hand-clap with a friend while holding a bag of crisps. 'Put them in your pocket,' said the friend. She didn't; and she managed to go through the motions of a clapping chant new to this playground in which the tummy must be rubbed, the nose held, a baby rocked, and a pistol fired:

My mother is a baker,
Yum yum ee, yum yum ee.
My father is a dustman,
Yum yum ee, yum yum ee,

Poo poo ee, poo poo ee.

My sister is a darling
Yum yum ee, yum yum ee,
Poo poo ee, poo poo ee,
Cushie coo ee, cushie coo ee.

My brother is a cowboy,
Yum yum ee, yum yum ee,
Poo poo ee, poo poo ee,
Cushie coo ee, cushie coo ee,
Bang![1]

They twizzled round and shot each other and collapsed ecstatically into each other's arms. 'Sharon my sister taught it me,' said the redhead. 'Have a crisp?'

The editor of *The Form Three Times* halted beside me, saying, 'Can you spare a minute?' He obviously had some hot news to deliver, and he is so mature-looking, so heavily bespectacled, it was difficult to believe the news was not of the utmost importance. 'First of all,' he said, looking over his shoulder. 'First of all we saw two of our friends who were sort of spying on us—' He shouted to another boy, 'Leslie, come over here!' He could hardly spare the time to convey this dispatch from the front line. 'Yes, well, first of all,' he dictated—'oh, you've got that bit. Well, so we started chasing *them*. Oh, here they come.' He pelted off. With a little imagination I could understand that the situation was at least as dangerous as that in Afghanistan.

Other parts of the playground were more static. Two girls stood with skipping ropes and skipped as many skips as they could. 'Not many people are skipping, are they?' I commented, when they paused for breath. 'I don't know,' said the bigger one. 'I only brought my rope because my friend did.' I saw a small girl struggling along with one foot in a Rota-Skip. She had no idea how to make it work. 'I forget what that thing is called,' I said. 'I don't know,' said the defeated owner. 'I got it when I went to visit Father Christmas in Portsmouth. He gave it to me.' Father Christmas must have bought up some old stock, the wily old bird.

[1] See I. and P. Opie, *The Singing Game*, 476–7. The girls in this school seem to think of the sister as being a baby, whereas usually she is grown up and is a 'modeller', hairdresser, or beautician.

A girl of a Peake-like paleness, with a deep slow voice and long straight hair, fixed me with a tragic look. 'Do you know what I would be doing if they let me out of this playground? I would be riding.' I wrote this down. 'Why do you write everything down?' she asked. 'Is it your hobby?' 'Yes,' I said. 'I will tell you what my hobbies are,' she said. 'I collect rubbers, and stamps, and I have a mouse collection—toy mice and cuddly mice.' 'I collect games, mostly,' I said. 'Oh, I can tell you some games we play,' she said heavily. 'There's "Please, Mr Crocodile, may I cross your golden river in your golden boat?" and "What time is it, Mr Wolf?"' 'Thank you,' I said.

The fourth-years are so tall nowadays they almost unnerve me. They seemed to have grown even taller during the half-term break. I found myself talking to these 10- and 11-year-olds as though they were too old to play games, although it is not entirely true. 'Marbles are not much in evidence this morning,' I observed. 'There are a few people playing,' replied an elegant dilettante, 'but the marbles season has gone out.' 'It'll be the football cards season next, won't it?' he said to the mouse boy. 'They get those cards with football players on them—Kevin Keegan and that—and throw them.' 'But how do you *know* it will be the football cards season next?' I asked. 'Well, it was last year and the year before that and the year before that. We fourth-years, we remember. The marbles season starts when we get a new load of first-years [i.e. in September]. And when they grow up they move on to football cards.' 'But I've seen fourth-years playing marbles like anything,' I protested. 'I know,' he said, 'but they're just mucking about.' I have learned that the more confident the statement, the less likely it is to be true.

The mouse boy and his boon companion came running back. The mouse boy said, 'I'm Zebedee, and 'e's Florence, because we're playing a game.[1] Well, you see, when 'e calls my name I have to run away and 'e has to catch me and when I'm caught I have to do the same.' They were thrilled with their invention. 'And we've got another game as well,' they said. 'We've got

[1] Zebedee is the small person on a spring, and Florence is the little girl, in Serge Danot's famous cartoon TV series *The Magic Roundabout*.

this game called Abbreviations. You think of words. Well, I'll think of one. Well, I've thought of "Orrible"—you miss out the "h", you see—and "Ead".' 'And then what happens?' I asked. 'Nothing. You can have, like, OBE. Well, I don't know what that stands for but *we* say it stands for "Old Big 'Ead".'

I could not raise any news from Lisa and some other old acquaintances, now 10 or 11 years old. Lisa is maturing, and no longer wants me to watch her invented games and dances. During our last two encounters she was more interested in showing me her new clothes.

The hand-clappers were still clapping when the bell rang. With half an ear I had overheard them singing the favourite Brownie song 'I went to the animal fair, The birds and the beasts were there', and chanting the newest version of 'When Susie was a baby', in which the schoolgirl Susie thrusts a hand in the air and screams 'Miss, Miss, I want to go to piss.'[1]

Wednesday 5 *March*

Boisterous March has arrived at last, and is whipping the land into activity: ploughing has begun, and cement-mixers are turning again in the half-built housing estate near the river. The stationer's window is filled with Mother's Day cards, most of them showing Victorian children picking primroses and violets in country lanes.

Two whirling dervishes were putting themselves into a trance outside the main door. They were circling with their eyes shut and arms held out and were not to be interrupted. Only when they had staggered to the fence and were hanging there, exhausted, could they say what the whirling had been like. The dark boy said, 'You get giddy and you feel sick.' The fair boy added, 'If you lie on the ground afterwards and close your eyes it makes you feel as if you are rolling downhill.' The boys who stood admiring them nodded and said it was true.

A row of girls lay on the ground wrapped in their overcoats,

[1] I. and P. Opie, *The Singing Game*, 458–61.

refugees from the schoolroom ('We're not allowed to stay in the classroom during playtime.'). The tall dilettante from last week approached them carrying a clip-board. 'I want you to write your names,' he said. 'We're going to send them to the BBC.' They knelt and added their names to the column. The dilettante's companion said to him, 'Usually when they have a vote they have more than a hundred names. Are you sure you've got enough paper?' The dilettante did not reply. He said to me, 'We're trying to get the whole school. It's about a film series called *Grange Hill*. It's to say whether you like it or not.' 'It's about a comprehension school,' said his helper, 'and they use bad language and that.' 'Nobody asked us to send up the names,' said the dilettante, 'but yesterday there was a programme of adults and children discussing about this series, because it's got—well, they talk to the teachers in rude language, and this character Alan, he smokes and people might copy him.'

'Must go and get some more games,' I muttered. 'Oh, can we come with you?' asked the girls. It was as good as a shopping expedition. They danced round me as I went towards the back of the playground. 'Those lot, they're playing Fairies and Witches. You have to stand round in a ring and everybody says, "Fairies, Witches, Fairies, Witches," and a person swings the rope round and you jump over it and if you get tangled up when it's "Witches" you have to take the end.' Then they flung themselves into a demonstration of what they now call Round the World in Eighty Days. 'I'll be on it.' 'All right, Mandy, you be on it.' The twizzler twizzled everybody in a flash. They explained as they went along. 'Mandy has to say what you are.' Mandy went round saying, 'You're a ballet dancer,' 'You're in a museum,' 'You're praying to God,' 'You're a frog.' 'Then we all have to be whatever we are,' they said, and performed accordingly. 'And then she chooses the best.' 'And then the best crouches down. And then the others dance round her singing, "Wakey, wakey, sleepy head, Don't forget to make your bed."' 'It's funny,' I said, 'I've seen all that before, and it still seems new.' 'It always is,' they said.

The teachers were arriving, and the children shrieked and shouted in their urgency to tell me the rest of the games before they had to go in. 'No, marbles is *not* in. They used to play marbles but they don't now. They play cars instead.' 'We play cars all the time,' said a tough operator, running his toy car over his neighbour's head. 'We play with the dinner ladies at dinner break, Mrs —— especially. She points to us when she sees us and you have to go and hide somewhere else, behind the dustbins or behind the loos.' They became incoherent describing the joys of hiding from the dinner ladies. 'I think you'd better go in now,' I said.

Tuesday 11 *March*

The children were late coming out. I went to pass the time of day with the caretaker, who was filling a plastic can from the tank of heating oil. 'Funny day,' he said. It was. A corrugated layer of cloud prevented the sun from coming through, and the trapped light was dazzlingly white. 'Been a good winter for heating the school?' I asked. 'Had a bit of trouble with the coke heaters,' he said, 'but yes, we must have saved a lot of fuel.'

The doors of the canteen were flung open and the children filed out. I walked with them. 'You sit on the floor during Assembly, I expect,' I said to the nearest boy. 'Yes, the fourth-years stand at the back, the third-years kneel, and the first and second-years sit.' 'Assembly is in the canteen, isn't it, where you have meals? Do *you* have the school dinners, or do you bring sandwiches?' It proved a popular topic of conversation. 'I bring sandwiches,' he said. 'I don't like the school meals.' 'What do you have in your sandwiches—ham, corned beef?' 'No, I don't like those, and I don't like cheese. I don't like brown bread. I sometimes have granary bread 'cos my Dad makes it. But usually I have white bread. I have chocolate spread, or tunny, or chicken—that's my best. Did you hear about the blind carpenter? He picked up his hammer and saw.' He ran off to join the footballers.

The lorry driver took his place. 'I got a good joke,' he announced. 'I just turned on the telly last night and I heard this joke. There was three men waiting outside the pearly gates and

St Peter says, "Right, first person?" and the man said, "I was suspecting my wife was having an affair and one day I went home and saw the bedroom curtains closed. I went into the bedroom and there was my wife sitting on the bed in a see-through négligé and I said, 'Where is 'e?' So I searched the flat and she says, 'Who?'"' He paused for breath. A bystander said, 'I heard that one, too. It's good.' The lorry driver continued, 'Well, it goes on—um—"Then I heard the lift go down to the bottom of the flats and I just rushed into the kitchen and looked out the window and saw the milkman rushing home. So I picked up the first thing that was near me and it was the fridge and I threw it out the window and hit the milkman on the head." And St Peter said, "Downstairs! Next, please." So 'e saw the next man and the next man was the milkman and 'e said, "Well, I was just come out this block of flats and I was just going home to my lunch and this fridge hits me on the top of the head," and St Peter said, "Right, indoors. Next, please." Then St Peter says, "Right—what happened to you?" and the man says, "Well, it's like this—I was sitting in this fridge . . ."' There was a murmur of appreciation as everyone realized the cleverness of the joke; and I thanked the lorry driver sincerely, for he had dictated it piece by piece for me—not an easy job.

The games seemed stunted this morning. Each one was somehow wrong. The boys playing War stuttered, 'Do, do, do, do,' from behind the dustbins, and no one fell down. Three boys played Fairies and Witches clumsily with a skipping rope, and then tried unsuccessfully to skip in an orthodox fashion. A game of Sly Fox used the marbles fence as 'home', proving beyond doubt that the marbles season is over.

Two boys in stiff new green anoraks wanted to tell me jokes. 'What is yellow and goes in a washing machine? A banana skin.' I had to remind myself that bad jokes are as interesting as good jokes. The boy—the elder of these two brothers—had had difficulty in pronouncing 'washing machine'. I thought I should let him go on with his jokes, if only as therapy, but his brother's need was just as compelling. 'Why—did—the—chicken—cross—the—road? To get to—the other end.' 'Where do you live?' I asked them, and 'How old are you?' (I felt I needed some sociological details.) They gave me their whole address, on one of the older council estates, and said they were 8 years old and '7—nearly 8'. 'We've got lots of jokes,' they said. 'Doctor, doctor, I keep thinking I'm a bridge. What's going over you? [meaning 'What's come over you?'] Two cars, two motor bikes, and a bicycle.' 'Why did the boy take a ladder to school? Because he wanted to get to High School.' 'What did the big candle say to the little candle? I'm going out tonight.' The jokes were being ground out of them. It was as if two rusty machines were working in slow motion; the delivery was entirely mechanical, and an achievement in itself. They did not look at me, far less look for any response; and I could not have stopped them if I had tried. Other children circled round saying, 'I gave him that one *ages* ago.' I plucked up the courage to say thank you and walked away.

On the far side of the playground a small boy was being dragged along on his back. When I reached him he was on his own, rubbing his ankle and crying a little. I sat down beside him and asked, 'Why were they doing that?' 'Jason pretended it was "tens"—'cos we were playing Om Pom—and it wasn't "tens" we were counting in, it was "fives"—and then all those boys came and fighted me, so I fell down and they pulled me along.' He was nearly in tears again. 'Well never mind, here's your story all written down,' I said, tapping my notebook. His misery subsided immediately. 'Oh *yes*,' he said, smiling, getting up and running off to his classroom.

Thursday 20 *March*

A savage wind straight from Russia. The sun was shining bravely.

The two boys in new green anoraks descended on me as soon as I entered the playground. 'I got a joke!' said the elder. 'What's yellow and goes in a washing machine?' 'A banana,' I said automatically, wondering why he was asking me the same riddle he asked me last week. 'A banana *skin*, not a banana,' he said scornfully. 'And I got another one. Doctor, doctor, I feel like a bridge . . .' I looked at him hard. Had he no recollection of telling those same jokes last week? Did he think I would want to write them down again? Or was he testing me to see if I remembered the answers? His younger brother joined in. 'I got another "Doctor, doctor" joke. Doctor, doctor, I feel like a tennis ball. I'll play with you later.' At least it was new, though not at all clever. Probably he had invented it himself.

I walked over to the grass verge where the boys were playing with toy cars. Certainly cars are still the rage. Stylish limousines were being given the treatment they receive in Roger Corman's films. They were falling over precipices, crashing head-on, going out of control and landing on their backs. 'Nobody's playing football cards,' I remarked, wondering whether the dilettante's confident prediction of the week before last would come true. 'No,' said several of the car drivers. 'There's a few people playing marbles over there though.'

The benches by the main door were as jam-packed with children as if it was summer. The boys on the first bench were playing with football cards, which come from bubblegum packets. 'That's funny,' I said, 'I was just asking about football cards.' 'We're a League,' they exulted. 'We've just started a League.' They were not skimming the cards as predicted by the dilettante, they were playing a game like Snap. On the other bench the tearful victim from last week was hunched over an exercise book, his feet sticking out over the edge. Another little boy was jigging about in front of him. 'He's trying to kick my foot,' said the victim placidly, 'and I am drawing a house.'

A few people were indeed playing marbles, though not enthusiastically. A girl player was more than willing to give me an interview. 'Yes, at the moment not many people are playing. There aren't many playing in the winter. When the summer starts—when the sun comes out a lot—when it's really hot—that's when we start playing marbles.' Substitute 'spring' for 'summer' and that seems about right.

I saw a little spindly girl bending over a drain cover. She jerked upright, moved round, and bent again. 'Ah, a loner,' I thought. 'I haven't met one for a long time, I'll go and have a word.' I followed her into the angle of the wall by the front door, where she was immediately absorbed into a vortex of excited femininity. A child in blue-rimmed spectacles shrilled, 'I'm the witch! She's been tidying up the house for me. I have to send them all out to work for me. *That* one was mending the road, and *that* one was cleaning the stables.' 'What if they don't do the work properly?' Her eyes gleamed. 'I put 'em back here in the dungeon, don't I? This is my cave, where you are. We're playing Fairies and Witches.'

In the corner by the S-shaped diagram girls were whirling with their arms out, letting the wind take them. Two girls were walking heel-and-toe from opposite ends of the diagram. Girls were twizzling, hand-clapping, jumping in the air and landing on the ground cross-legged ('It keeps us warm'). When the bell rang we all walked to the classroom together, and one of them said, 'You know that other game called Fairies and Witches, with a rope? Well, it's been stopped because someone got hurt—we don't know who.'

Tuesday 25 *March*

The weather is striving to become normal for the time of year. The wind has veered south. A deep blue sky was almost covered in towering black and white clouds, between which the sun intermittently shone. The stationer's window was full of Easter cards that used every conceivable combination of Easter symbols: rabbits, crocuses, churches, crosses, apple blossom, baby chickens, primroses, daffodils, lambs, and Easter eggs.

Shrill screams reminded me I ought to be in the playground. By the time I arrived the screamers had disappeared (screamers are usually running) and the only sound was 'De-de-de-de-de' from the Olympians pushing their cars.

Three little girls skipped towards me in a row: one held a knitted doll, one held a fluffy bunny, and the other held a bag of Noughts and Crosses crisps. They stood with me, companionably. An older girl ran up. She looked at me with sparkling eyes from beneath her fringe and said winsomely, 'Can I tell you a joke?' 'Go ahead,' I said.

'I got this from my sister,' she said. 'She's 13 going on 14, and she goes to the convent school. You see, there was this vicar and he was riding along on his bike and another vicar was walking along on the pavement. And the one on the bike said to the other one—I've forgotten their names—"Where's your boicycle, vicar?"' ('Oh, was he Irish?' I said, not sure of the accent. 'It's Irish or some kind of accent,' she said, 'it's a very old story.') 'The other vicar said, "Someone stole it. I didn't see them steal it, but it's gone." "Why don't you give them all the Testament on Sunday, it might shake someone's brain and they might bring it back." So he did that on Sunday and the next day they were both riding along on their bicycles and the vicar passed the other vicar and said, "Oh good, I see it worked." And the other vicar said, "Oh yes, I got to the eighteenth chapter and I remembered where I had left my bicycle."' I did not comment on the story. She had told it with such brilliant fluency it would have been churlish to say that the last time I had heard it, many years ago in adult society, the vicar who had mislaid his bicycle tried reading the Ten Commandments in church and remembered where the bicycle was when he came to the seventh commandment.

'I was sent a rhyme this morning from Scotland,' I told them, 'something about love and marriage and a baby carriage.' The girl looked at her neighbour and giggled. Then she started singing, to the tune of the 'Campdown Races',

> First comes love, and then comes marriage,
> Then comes Rachel in a baby carriage.

('*In* a baby carriage?' 'Yes, in a baby carriage.') 'It's for when people are in love,' she said. 'We sing it in the classroom when the teacher's not there—to tease them.'

I looked over their heads towards the marbles fence. 'Quite a lot of people playing marbles,' I observed. 'Why's that?' 'We're not allowed to play with balls,' they said. 'Not football, not Donkey, not anything. 'Cos somebody broke a window. We're not allowed to play Piggy-Backs, 'cos this boy broke his wrist. We're not allowed to play Trains either, because people hurt other people.'

I walked along the lines of 7-year-olds who were waiting for their escort. 'Any games?' I said. 'What have people been playing?' 'We've been hopping on the circles, trying to catch each other.' 'We've been playing Exercises, like in PE,' said the witch and her fairies from last week. 'We've been Piggy-Back Fighting,' said two female imps, who were astonished to hear that somebody had said it was not allowed any more.

A coffee-coloured cherub and his stout pink friend were dancing about chanting a rude rhyme when the teacher arrived. I walked over the road with them and asked the teacher if I could borrow just those two. ''They're Miss ——'s,' she answered, 'but she won't mind.' 'I didn't get all the words right,' I said to the cherub (beautiful dark liquid eyes, he had, and curly hair). 'Well,' he said, 'it goes,

Milk, milk, lemonade,
Round the corner chocolate's made,'

and he stabbed a finger (rather less confidently than before) at his breasts, crotch, and bum. 'Thank you,' I said. 'Now let me see, how old are you?' 'He's 8 in May,' said his friend, 'and so am I.' 'There's another one I know,' said the cherub:

Tarzan swings,
Tarzan falls,

here he was convulsed with laughter—

Tarzan 'urts his 'airy balls.

'Better go into school now,' I said.

Tuesday 1 *April*

Sullen, threatening clouds, streaked with yellow.

As I entered the playground a teacher called out cheerily 'You'll have to be careful today—April the First, you know.' When the first little boys ran up with jokes I said, 'Miss —— says I've got to be careful.' Straight away one of them said, 'Look at that ball up on the roof!' and straight away I looked up at the roof—so slow am I to take a warning. 'Your shoe-lace is undone,' said another; but I could not be taken in by that old chestnut.

'We've had our teachers on,' they said with gusto. 'The *best* thing is to turn the teacher's desk round so that the drawers are the wrong way.' 'Yeh, and when she was reading *Signpost*—it's a horrible book—me and Dennis and Nicholas hided all her books and then she said, "Where are my books?" and she looked everywhere—into the stock cupboard and everywhere—and then we told her.' The girls came bubbling up with, 'We haven't done ours yet. All the girls are going to hide. We're going to put all our chairs up on the desks and pretend we've gone home, but we're going to hide in the toilets and then we'll come out and say "April Fool's Day!"'

A small girl hooked her arm into mine. 'I'll tell you what games people are playing,' she said. 'There's lots of people playing marbles. And there's people skipping, behind the fourth-year block. And *we're* playing "Teddy and me went out to tea". We'll show you.' She and her friend (both 8 years old) crossed their arms and clasped hands and danced away from me chanting, 'Teddy and me went out to tea, Locked the door and turned the key.' They took infinite care about starting on their right feet. They took a step, hop, and step to the right, then a step, hop, and step to the left, and continued like this to the end of the rhyme, when they reversed with great neatness, still holding hands, and set off in the other direction. When they got back the smaller one said 'Turning is the hardest part. You have to practise. I couldn't do it when I was 7.' 'I could,' said her friend.

We went to watch the skipping. It didn't amount to much. I heard the alphabet being chanted, and it was nice to hear chanting of any kind (it has the sound of life reviving); but the other skipping games were solitary affairs.

The boys had pursued me and one of them said, 'You've got something on your shoulder!' 'I'm not being caught again,' I said, looking him in the eye. 'It's true. What is it, let me see? It's a spider!' He picked a black plastic spider off my shoulder and held it up. Everybody fell about laughing. I was glad I was with them for April Fool's Day; such hilarity is not to be had in adult society.

A boy with a tousled mop of hair and naughty eyes said, 'Mrs Opie, I've thought of a good game—come and see.' I sat down beside him at the foot of the classroom wall. 'You get an old cover of a pen, in good condition like this.' He waved the outer cover of a biro, then put the end in his mouth. I thought perhaps he was going to smoke. 'Then you put a bit of a rubber in it, and you blow, and you fire it up a girl's skirt.' The girls flapped their skirts and screamed. They were saved by the bell.

A young teacher in black cord trousers and a white fisherman's sweater greeted me from across the playground. She was holding something at arm's length. As she came nearer I saw it was a soft plastic centipede with a dragon's head. 'In my handbag!' she laughed as she passed on her way to her classroom. 'Don't forget the term finishes tomorrow.'

Tuesday 15 *April*

Summer term began yesterday. It is more like May than April, with temperatures in the 70s. The sky was pastel blue, today, chalked with round white clouds, and the wooded hills in the distance looked soft and inviting.

Several small boys gambolled up and stood in front of me. They wanted to be friendly but they did not know what to say. 'Did the holidays seem long or short?' I asked them. 'Short!' they chorused, as expected. Conversation languished again. At last one of the boys said, in a despairing voice, 'I *can't* think of a joke.' They scrimmaged a little (which always clears the brain) and one of them came back to ask, 'Why did the elephant cross the road? To go to the zoo.' 'I just thought that up,' he said, and told it to a girl who happened to be passing.

I had seen a row of girls skipping across the playground each in her own single rope. I thought I should leave the 7-year-olds and look for some games, but the 7-year-olds had become chatty and possessive. 'Did you know Blondie died yesterday? She was hanging on the telephone.'[1] 'Did you know Tommy Cooper died? He died "just like that".' ("Just like that" is the comedian Tommy Cooper's catch phrase.) 'This seems to be a

[1] Refers to the pop group Blondie and their song 'Hangin' on the Telephone'.

new kind of joke, with different people dying in different ways,' I said, and they all looked very pleased. We set off round the playground together.

The skippers had disappeared. A few boys were playing marbles, and the football area and side annexe were full of cheering footballers. ('If you want to know what's going on, it's *noise*,' said a senior boy in passing.)

Four 10-year-old girls were giving serious attention to a game I did not recognize. They were standing on a line of the netball pitch, jumping forward, jumping forward again, and then marching forward a certain number of steps. Because it had this definite form, and a song to go with it, the performance looked important and interesting. (Perhaps there is an equation: form plus rhythm plus seriousness equals ritual, which equals importance.) The girls, however, would not say what they were doing. They giggled and giggled. Even when I showed them how few games I had collected this morning they would not stop their nervous giggling. 'But ours isn't a game at all,' they said. 'It's only a song, with actions.' 'What song?' 'It was in *The Sound of Music*.' 'Oh, I know, Julie Andrews was in it.' 'Yes, it's one of her songs,

> Doh, a deer, a female deer,
> Ray, a drop of golden sun . . .

We'll show you if you like.' But they did so shamefacedly, like adults who have been caught larking.

The 7-year-olds felt neglected. 'Look, we got a game,' cried one urchin, and launched himself at a friend. They tumbled and wrestled on the asphalt, their trouser seats dustier each time they gained the mastery. The girls looked down at them and laughed. All at once the challenger leapt to his feet and presented the nearest girl with half a glass marble.

The girl looked at the piece of glass marble and laughed again and put it in her pocket. She turned to a girl friend, buttoned their two cardigans together, took the other girl's hands and began twizzling. They twizzled away into the distance. Soon the rest of the group dispersed, leaving me only with the dependent child in blue-rimmed spectacles, who was dragging on my arm and sucking her thumb. I longed to ask her about her home background, but it seemed impertinent. We walked towards the alley-way by the toilets, where two games of Sly Fox were in progress. We noticed the singing game Orange Balls on the way, which changed to Round the World in Eighty Days as we went by.

One of the grown-up boys, nearly 11, came out of the toilets and said to me, 'Oh, hello! Have a good Easter? D'you know about the Irishman and the fish-and-chip shop? An Irishman went into a fish-and-chip shop and bought some fish and chips. Ten minutes later he came back and said to the man, "Are you sure this is fish you sold me?" and the man said, "Yes, of course. Why?" and he said, "Well, it's just eaten my chips."' 'Lovely,' I said, and began to move on. 'Hang on, hang on, here's another one,' he said. 'It's about an Englishman and an Irishman, and they were in a jungle together, and a great live lion was chasing after them and the Englishman shouts, "Paddy! run, run!" and Paddy says, "Not me—it was you who threw the stone."'

The seniors fairly whisked inside when the bell rang, but the 8-year-olds still milled around outside their classrooms. The William boy brought a friend forward, saying, ''E's got a joke for you, I think.' The fair-haired friend twisted a lock of hair and said doubtfully, 'Do you want one?' I nodded enthusiastically. 'Well, there was this business man, you see, and he was riding on a motor scooter, and he wasn't wearing a crash helmet. And this policeman stopped him and said, "Why aren't you wearing a crash helmet?" and the business man said, "I didn't want you to think I was a Mod."' ['Mods', gangs of youths on motor scooters who fight gangs of 'Rockers', have been in the news again recently. They wreaked havoc in Scarborough and other seaside towns on Easter Monday.]

Wednesday 23 *April*

Cool, grey, astringent; almost a relief after the Disney-pretty spring days we have been having.

A small tough in a poplin jacket with a furry collar accosted me over the fence. 'Say, "Bake me." Go on, you must say "Bake me."' I said, 'Bake me.' He replied, 'I can't, I haven't got the recipe,' hurled his toy car across the playground and bounded after it. When he came back I asked him, 'What kind of a car is it?' 'Ford Three,' he said nonchalantly. I looked at the car's belly and read out, 'Ford Capri II'. 'Oh yeh,' he said, 'and slightly scratched.'

The tall blond boy with laughing eyes (a playboy if there ever was one) came and opened the gate for me. 'There was a brown sauce and a red sauce, and the red sauce said, "I can't ketch up",' he said, spluttering, and tossed a long seductive look at me. I could scarcely squeeze into the playground for a skipping game that was blocking the entrance. Three droopy little girls were trying, ineffectually, to skip in a toy skipping rope with rag doll handles. 'They haven't even got faces,' I said despairingly. 'They *had*,' said the owner, 'both sides. One was a smiley face and one was a sad face—but they got kind of rubbed out.'

The playboy accompanied me to the marbles fence (where marbles, like the skipping this morning, were neither in season nor out). 'I know what you're going to say,' he declared. 'Well, I'm not going to say "What?"' I countered. 'You still said it,' he rejoined, giving me a ravishing glance. 'Where do you weigh whales?' he went on. 'I know that one,' I said, 'it comes from Sandra McCosh's *Target Book of Jokes*. The answer's "At a whale-way station."' Well, everyone's getting their jokes from that book now,' he said. 'Which jumps higher, Bert or Ernie? Neither. Trees can't jump.' 'But Bert and Ernie aren't trees,' I objected. 'Does it matter?' he said.

All this while I had felt a light tap at regular intervals on my back. Because the taps were light and regular I knew they were part of a game. The two rosy 8-year-olds knew I would not mind. They knew I would appreciate their amazing invention. 'We pretend to skip in a not-rope and then we flap you in between whiles.' It was certainly more fun than the few orthodox skipping games (chiefly Building Up Bricks), which were laborious and badly attended.

Today was a *dies non*: no entertainment, no games worth recording. The fifteen girls playing leap-frog were crammed so close together that the leaper could scarcely land between them.

The marbles games lacked excitement, and the game of Sly Fox was remarkable only for its extreme size—the chain stretched right across the playground.

Tuesday 29 *April*

I passed a knot of women chatting in the village street. A pleasant apple-cheeked woman was saying to the two others, 'And she said, "It's that cold this morning I'm not stopping to talk to *anybody*."' It was cold, too, with a restless grey sky and a biting wind.

Some girls came aeroplaning towards me and stopped in a companionable but speechless line. 'You were kept in Assembly a long time this morning,' I said, 'was it something special?' 'Well,' said one of them, considering the matter, 'Mr —— was telling us a story. He was reading this story about this boy, and his clothes were on fire.' 'A *long* story,' interrupted another. 'He went to hospital, and he had to have these brass things on his body where the fire had made holes.' 'Is it true?' I asked. 'I don't know,' she said impatiently. 'Anyway in the end he was back to normal but he still had marks where the holes had been.' 'Where the grafts had been,' corrected the second girl. 'Do you think it was to teach you something?' I asked. The first girl shrugged her shoulders. 'Shouldn't think so,' she said.

'Can I tell you something *really* interesting?' said a boy, pushing his head under the girls' arms. 'Don't take any notice of him,' pleaded the redhead who had been reporting the story. 'Michelle loves Keith Clarke!' he shouted, and ran away. 'It's my turn now,' said another boy. (The boys were in a hurry, they wanted to start their football.) He said, 'It's not a joke, it's a problem. There was this man who lived on the tenth floor of a flats and he only went to the fifth floor. Why?' I looked in the air and down again. 'I don't know.' ''Cos he was a dwarf and he could only reach the fifth floor button.'

The girls were glad when the boys went away. 'Now,' they said, 'can we show you something we made up?' They recited the well-known skipping chant:

I'm a little Girl Guide dressed in blue,
These are the actions I must do.
Salute to the Captain ['You have to salute.']
Curtsey to the Queen,
And show your knickers to the football team.

'It's for when we do handstands,' they said. 'You start off marching, you see, and swinging your arms. Then you do the saluting bit and cross your feet, and then you curtsey and go down on one knee, and then you do your handstand.' It was a neat and appropriate use of the old rhyme, the only drawback being that the football team were too busy to notice.

Football was maniacal this morning. It had spread half-way down the side of the playground, and intoxicated spectators were yelling, 'Bluff it, bluff it,' at players they did not want to score a goal. A small boy bent down to place the Wembley Mini Striker ball for a corner. The crowd thought him slow, and chanted

Why are we waiting?
We are suffocating,

and other more manly exhortations. The girls looked at them sideways. 'Don't *say* that, it's *rude*,' said a girl vehemently. 'What did he say, I couldn't hear?' I asked. 'Well, *I'm* not going to say it,' she replied. They all looked demure. The first girl sidled up to me and whispered, 'F-U-C. It is rude, isn't it?' 'It's one of the worst ones,' I agreed.

One of the football spectators who felt able to give me half his attention said he had seen part of the Arsenal v. Liverpool Cup Tie match last night, when the teams had drawn for the third time running. 'I watched about the first half-hour,' he said. 'Yeh, it was *magic*.'

The girls and I had retreated to a bench where things were quieter. I learned that swimming would start in May, and that £84 had been taken in the Jumble Sale last Saturday. Then I remembered that a journalist had been on the phone yesterday asking about secret languages, so I asked if they knew any. They became very animated. A little one said in an awed voice, 'My brother knows one. He knows *French*.' The redhead said, 'My next-door neighbour's Dad knows one. You write the alphabet forwards and then you write it back to front.' A dark-haired, interesting-looking girl said, 'I used to belong to a thing called the Puffin Club, and they had a Code.' A thin, pale girl said, 'My sister goes to shorthand classes.' The redhead said, 'The boys in our class make up their own little language—it's mostly rude words like the one I told you.'

Further towards the road I could see lines of leap-frog players, and also some people leap-frogging over bodies hunched on the ground. When the bell rang I went to enquire and they explained, 'It's for people who can't jump.'

Tuesday 6 May

A knockabout day, the wind blowing one way and then the other, and enough ice in the wind to make the villagers keep their geraniums indoors.

One of the 8-year-old boys ran towards me saying, 'Guess what we've been doing in the classroom. We've been cutting out paper and you fold it over and then you unfold it and glue it on the back and stick it on a piece of coloured paper and Mr —— is going to put it on the wall. It's pictures and shapes and things like that.'

More boys gathered round, shouting and leaping. 'Look up, look down, Your knickers are falling down,' said one to nobody in particular; then he sobered up and said, 'You can only say that to girls, really.'

Two girls rollicking along with arms entwined were, I thought, singing the familiar song 'Ten green bottles hanging on a wall'. As they drew near the words seemed less familiar. 'It's not ten green bottles,' they tittered, 'it's ten blocks of dynamite—and it's swearing, really.' They sang, 'Ten blocks of dynamite hanging on a wall, Ten blocks of dynamite hanging on a wall, And if one block of dynamite should accidentally fall, There'd be no blocks of dynamite and no bloody wall.' 'I'll write it down b–y, like that, and then it won't notice,' I promised. 'And anyway it's a proper word, like red blood,' said one of the singers. 'And anyway it's not our fault, it's going round the school—round and round and round.' She waved her finger round above her head and staggered beneath it, like Harry Lauder demonstrating the glorious instability of Glasgow.

They have a genius for clowning, those two. Is it because one is a Mauritian, and the other half Maori? 'We're not pessimists, OK?' they cried, garrotting each other; and then remembered more formal entertainments. 'Knock, knock, who's there? Justin. Just in case . . . just in time for a cuppa tea.' I invited them to help me look for games.

We stood on the edge of the football area. They pointed out 'Football played with netballs', and 'One lot of skipping with seven people doing it'. They noticed someone launching a cardboard biplane (manufactured), and said that a boy writhing in agony on the ground had cramp in his leg. But it was I who first saw the coloured spectacles. They had been issued by the Unigate milkman over the holiday weekend. They were made of white cardboard with 'Be A Crazy Drinker' across the top, and the eye pieces were of red cellophane. 'They come with the Crazy Drinks,' explained the boys who were wearing them. 'Crazy Drinks come banana flavour, chocolate flavour, or strawberry. If you look at, like, her blue coat, it looks purple.'

'They're still doing handstands,' said the girls. We counted fourteen girls upside down against the west-facing wall. We skirted the football and found ourselves in between two large games of Sly Fox. One tiny game of marbles survived by the fence. The girls were, I thought, pushing reportage to its limits when they told me to write down Yellow Lines. 'You run along the yellow lines of the netball court and when you meet you bump into each other with your hands and shout, "Butlins!"' However, they assured me they often play it.

After the bell rang some of the senior girls came and said breathlessly, 'We've been looking for you. We've got some games to show you.' One of them had been given a plastic pot of Slymuck for her birthday. It looked like jellified orangeade. It was cool, and wet-feeling, and when held up it sagged but did not break. 'It smells sort of hospitally, like disinfectant,' said one of the girls, 'and it doesn't stick to your hands. It's just come out in England. The real Slime that you can buy, it smells horrible, and the mothers don't like it because it sticks to the furniture.'

The other game they wanted to show me was a card game called Junior Quartet, made by Waddingtons. The cards were miniature, and bore pictures of animals. 'It's quite good,' said the earnest owner, 'but it's better if you play this different game with it, that we just made up. It's called Punishment. If you pick a cat or a dog you get a punishment, like "Run up to there and back, and when you come back I'll murder you," or "Stand on one leg while I count forty," or "Give me a piggy-back."'

Tuesday 13 *May*

A really hot day. I warily cast a clout, and then found the steering wheel of the car too hot to touch. Deep summer reclined in the ditch beside the school: bluebells and Queen Anne's lace and tall nettles, with broad dock leaves reassuringly at hand.

Strains of West Ham's supporters' song, 'I'm forever blowing bubbles', came from the far side of the playground. Then I heard 'Ready, steady, go!' Then a chanting threesome goose-stepped towards me, arms round each other's shoulders,

> We walk straight,
> So you'd better get out of the way—
> Kick!

They did not carry out their threat. Instead they jumped up and down saying, 'We've got a joke. If the plural of hippopotamus is hippopotami, what's the plural of what-a-clotamus?' (They literally said it together. They helped each other get it right.) 'It's "What-a-clot-am-I", isn't it?' I said appreciatively.

The playboy had been hanging about. 'I'll tell you what my brother made up. I've got this very ancient brother—he's 5. It's this. "What's ancient and uses Germolene?" The answer's "A tyranno-sore-arse."' 'He couldn't have made it up,' I protested. 'He *did*,' said the playboy.

Girls crowded round, offering crisps. I could have a 'Big B Kwika Snak', or a 'Hula Hoop', or a 'Monster Munch'. They saw that I was too hot to move, so they told me what games were going on round the corner. 'They're not skipping round there, they're all walking on their hands. That's the craze now. You can see four games of marbles, can't you, but they're only mucking about.'

It is better, in hot weather, simply to gossip. One of them said, 'We've got a friend—well, not really a friend because she's a bit awful at times. She's always ill when it's something she doesn't like, and she gets well when it's PE [Physical Education] or swimming. So we say to her, "When God gave out the illnesses Ann was at the front of the queue."' Another girl added, 'She went for an audition for the Youth Orchestra, too, and when she came back she boasted. She thinks she's got a place. She went for an audition before, and didn't get in.'

Self-righteousness is a pleasant exercise. The redhead had her own theme. 'We're going to have a chart for people who've been naughty, in our class. Mrs—is just thinking how to do it, and we all made suggestions. 'Cos people *are* naughty. They're always gassing and throwing pencils. She thought she might make people pay a penny for being naughty and then use the money for a prize for the person who's goodest. That wasn't my suggestion. I thought they could stay in at breaktime and tidy the bookshelves.' 'That's not much of a torture,' said a hatchet-faced girl who had come up, 'better to make them eat the school dinners.'

Food proved an interesting subject, and nothing is so satisfying as proclaiming one's likes and dislikes. 'The dinners cost more, and they taste a *bit* better—not a lot.' 'There's not a choice any more. We get spaghetti out of tins—practically everything's out of tins—huge tins, you can see them, millions of them. But there's a lot of stuff thrown away in the pig bins, 'cos people don't like it.' 'You still get mushy peas,' said the hatchet-faced girl gloomily. 'I bring sandwiches. I make them myself.'

Monday 19 *May*

The ideal summer day: cloudless blue sky, light breeze, hot—but not too hot—sun.

I nearly fell over a pair of sandals as I went into the playground, discarded by a girl who was practising cartwheels. Acrobats were everywhere, and seldom the right way up to be interviewed. 'Cartwheels are too *easy*,' said a fairy-like creature, her eyes sparkling, 'what *I* do is arrow springs.' 'Arab springs,' said another crushingly. 'It *is* arrow springs,' insisted the fairy. 'You do a cartwheel and jump at the end and land facing the way you came. Hand springs is another thing you can do, but they are hard and I can't do them.' 'Why is everybody suddenly doing these things, though?' I enquired. 'Some people go to Gymnastics after school. Miss —— teaches them, in the gym. She's a gymnast.'

The redhead arrived and said, 'You want some more games, don't you? Those people over there are doing—what's it called?' (The children seem to know no more about games than I do—less in fact.) A rather flabby, moon-faced child was clapping, with a much smaller girl, in a soft but maddeningly efficient way. She was using the minimum effort to achieve the maximum effect, like a very feminine woman with tiny hands whose sewing is effortless. They were clapping to the Boy Scout camp fire nonsense song 'Ging gang gooly gooly watcha', with its satisfying 'Umpha, umpha, umpha' at the end ('You do the "umphas" straight,' they explained, meaning that they held their hands upright and clapped them straight on to their partner's).[1]

Playtime was over. The smaller boys retrieved their toy cars from the dustbowl in the grass fringe, and tipped the earth out of them.

[1] See *The Gilwell Camp Fire Book* (1957), 37.

Tuesday 3 June

A grey day, but warm with a muggy, moist warmth.

A heavy girl (her legs 'beef to the heels', as the Irish say) lumped across to the fence. 'I'm looking out for my brother,' she said; but when the 7-year-olds came over the road she did not speak to any of them. Perhaps she wanted to check that her brother had not absconded, or that the teacher had correctly supervised the crossing, or simply wanted a comforting glimpse of her brother—or he of her. Anyway, she immediately joined two friends who were sitting propped up against the fence eating their crisps.

I had nearly walked right through one of the 8-year-olds' nebulous pretending games before I realized it *was* a game. 'We are all playing Gymnasts. We are all gymnasts and we are very good at it, and we live in this big house, and we are very rich.'

I counted five games of marbles, and saw some desultory games of football. Everybody else seemed to be flipping themselves upside down against the wall, or into the circles painted on the playground, or doing nothing. I hung around near the drinking fountain. Little boys came one after the other to fasten their mouths round the nozzle in the centre of the steel bowl and press the lever down. 'You put your mouth down *right round it*.' 'You can press the lever any way you *want*.' 'You're supposed to press it *down*.' Some of them rubbed the top of the nozzle quickly with a forefinger to clean it before they drank. One lingered a long time, fiddling with the bowl. 'It's loose,' he said. 'It leaks. I'm trying to straighten it. Yesterday the water made a big puddle and it ran right across the playground to *here*.'

Some girls were dancing to and fro chanting, 'One, two, three, four, Heel, toe.' I said helplessly to the nearest girl, 'I *must* get some proper games.' She glanced round quickly. Two smaller girls were clapping very carefully to 'I love sixpence, Jolly jolly sixpence'. 'Actually', she said, 'I know thousands of clapping games. I'll get Cheryl Gardner.' She came back hauling Cheryl with her and they started a marathon performance that lasted until well after the bell rang.

First they did the question and answer routine about the different sailors and their matching wives:

> Have you ever ever ever in your long-legged life
> Seen a long-legged sailor with a long-legged wife?
>
> No, I've never never never in my long-legged life
> Seen a long-legged sailor with a long-legged wife.

And so on through the 'knock-kneed' and the 'bow-legged' sailors and their respective wives.[1] They patiently showed me the movements. 'Every time you start a new bit you put your hands on your knees and then clap your own hands together—that's for "Have—you" and "No—I've", because they're slow. Then you go quicker and clap against the other person's right hand and your own hands together again, and the other person's left hand and your own again, and when you say "long" you hold your hands wide apart.'

They fetched two more girls and did 'See, see, my baby', and 'I went to a Chinese restaurant'; and then we had to ask Mrs —— if they could stay out of school until they had finished their *tour de force*. ('Yes, certainly,' she said, beaming.) They clapped 'Om pom pee kara lee kara leska', and racked their brains to think of other clapping songs—for they wanted to stay out as long as possible. They went all through 'Who stole the bread from the baker's shop?', the circular clapping game; and they would have continued with 'Ging gang gooly' if I had let them. 'What about "Zana, zana"?' asked one. 'That's just a hymn,' objected another. 'People *do* clap to it,' said the first, stoutly (I felt sure they did—'people' will clap to anything). Eventually they ran out of ideas and I said, 'You'd better go in.' As I went out I saw that the puddle from the drinking fountain had indeed spread half across the playground.

[1] For this and the next four clapping chants see I. and P. Opie, *The Singing Game*, 453–5, 474–5, 463–4, and 448.

Tuesday 10 June

A lid of grey cloud kept the damp air down, and the insects with it. The house-martins were swooping low over the play-ground like stunt pilots.

The children came roaring out, and the boys flung themselves into position on the football pitch. Twenty-five boys, at least, stood in serried ranks droning, 'Why are we waiting?' and when the footballs still did not arrive they changed to 'Hurry up, hurry up, hurry up, hurry up,' clapping their hands to the rhythm.

Most of the girls seemed to be walking on their hands, or standing on their heads or hands against the wall, like a row of tortured bodies with red faces and brown legs. I complained to two girls who were passing by, still eating their crisps, 'It's either football or handstands, and no other games for me to write about at all. You don't go on to the field at all, this playtime, do you, only at dinner time? What do people play then?' 'Football and handstands,' she said apologetically.

They were joined by other senior girls, and one middle-sized boy. 'Want a joke?' said the boy. (At least joke-telling is always in season.) 'How does an elephant get down from a tree? He stands on a leaf and waits for autumn.' 'Weedy!' shouted a tough guy in a safari shirt. 'I got a better one. There was this Englishman, Scotsman, and Chinaman going in this plane over Britain and the Englishman looks down and says, "There's a bit of England," and the Scotsman looks down and says, "There's a bit of Scotland," and then this Chinaman he feels in his pocket like this and he pulls out a cup and he chucks it out the plane and he says, "There's a bit of China."' The tough guy looked round at the others and said, 'Know that one?' and ran off to his classroom. The others realized that the bell had rung and ran off too.

Tuesday 17 June

There was no morning or dinnertime play. The rain had come down like Niagara. By 2.30 p.m. the clouds had gone, the sun was hot, and the children were making whoopee in the playground.

I stood looking at them for some time, wondering why the scene reminded me of the traditional curtain-raiser of a provincial pantomime—'Village boys and girls merry-making'. Of course it was because, unusually, the boys had joined the girls in a variety of peaceful games. Some were merely on the fringe, jumping over little girls who were crouching down, or playing a long game of Piggy in the Middle. But others had been caught up in the acrobatics craze, and were seriously attempting the handstands, cartwheels, and walking-on-hands that the girls have been busy with for weeks. Three of the boys who were trying to walk on their hands seemed embarrassed for a moment, but when they saw that I was not going to laugh they asked me to monitor their progress. 'I stayed up for three seconds then, didn't I?' said one. Although a pair of long grey flannel trousers upside down seemed novel, if not ludicrous, I remembered that it is the unaccustomed that seems comic, and that in the seventeenth century it was the boys not the girls who were practising such feats.

Over by the gate two 8-year-old girls were engaged in a vigorous and unfamiliar little performance, with the words:

> I think Chinese people are funny,
> This is how they spend their money;
> Oom cha, oom cha,
> Turn around, boom cha.

They began by moving their arms up and down like guardsmen clashing cymbals. 'Then when we go, "Oom cha,"' they said, 'we fold our arms and bang into the other person, and when we go, "Turn around boom cha," we turn around and go boom into each other with our two bottoms.' They were disconcerted when I asked where the game came from. They consulted each other and said doubtfully, 'We think Jackie's little sister made it up. She goes to the Infants.' I hardly recognized it. For about thirty years it has been a popular skipping rhyme in a form like this:

> German boys are so funny,
> This is the way they earn their money,
> Oroompa-pa, oroompa-pa,
> Oroompa-pa-pa-pa-pa-pa.

Although there is no direct evidence, this must surely have been a rhyme chanted to annoy the itinerant German bands at the turn of the century.

A boy had been cannonading into us. I rounded on him when we had finished our business, and told him to shut up and go away. He looked surprised. I asked the girls what they brought for their dinners. They perked up tremendously. 'Packets of crisps,' they said lovingly. 'Apples, pears, bananas, and yoghurts—usually yoghurts—and sometimes jellies. We bring them in pots.'

More boys were playing marbles than I had seen for some time. A large game at the end of the fence suddenly broke up, and the boys ran away. I asked the only remaining boy what had happened. 'They're frightened of Derek,' he said. 'Derek just broke it up. Derek's got a joke to tell you.' He nabbed Derek—who was of course the human cannon ball of the previous incident—sobered him up by pushing him into a puddle, and told him to tell his joke. 'I've forgotten it,' mumbled Derek. 'It was something about a fly getting squashed,' said his friend.

Wednesday 25 *June*

Brilliant sunshine for the school's annual fête in aid of charity, which lasts a whole week during morning playtime. This year's was 'in aid of East Africa where the children are starving'.

I was greeted from afar by Mrs ——. 'Not many games going on,' she called. 'No,' I said, 'all I've got written down so far is,

"Everybody eating cakes."' 'That's right,' she said, looking down at a flapjack she was eating, as if for confirmation. 'Do you fancy this?' she asked, holding out a pink iced fairy cake. 'I've just bought it.' 'Ooh, thank you,' I said, 'what a treat!'

I made for the first stall, 'Guessing how many marbles in a large Nescafé jar'. The guesser was taking it very seriously, bending to eye level to assess the number of layers and number in each layer. She was, however, eating a fairy cake at the same time. A blonde 9-year-old bounced up to her and said fiercely, 'Those cakes are lovely. Do you realise that?' I was not sure of the motive behind this remark.

The next sideshow was a competition to get a marble into the hole of an upturned flowerpot by lifting it from a box with a fork. 'Come and Try Your Luck,' said the placard, 'Only 2p. Get the Marble in the Flowerpot and Win 3p.' (None of the placards mentioned skill, I noticed, only luck.) The current competitor was having no luck at all, but I was told that four or five people did it yesterday.

Round the corner a long queue of boys was waiting to throw bean bags at a wall of tin cans. 'Double Your Money,' said the wall poster. 'It's 2p for three goes,' said a maiden in crisp red and white gingham, taking the money from the boys as their turn came. Her friend sat beside her with a small pink plastic handbag hanging daintily from an outstretched finger, stashing the money away and licking a toffee apple. This was the most popular sideshow, and produced much cheering.

Along the grass verge quieter sideshows—'Guess the Weight of the Hamster', 'Guess How Many Raisins in the Cake'—culminated in a drinks stall with one rather flustered 10-year-old in charge. 'Come and Try Our Thirst Quenching Drinks,' said the advertisement. 'Now—don't push—line up *please*,' pleaded the saleslady.

A friend of mine was getting little custom with his 'Roll Up, Roll Up, Try Your Luck on the Exciting Game of Roulette Only 2p Ago' [*sic*]; but the Aunt Sally game was doing steady business. A succession of cheerful faces looked through the hole in the cardboard and were seldom hit by the water-laden sponge. 'There are four of us and we take turns looking through,' said the stallholder. 'The teachers are going to have a go later on.'

The roulette stallholder's business was so slack he helped me report on the rest of the stalls. 'There's a Knobbly Knees competition, and there's moving dried peas from one plate to another by sucking with straws.' I added, 'The old lottery game, with rolled-up numbers inside half straws, and the bucket nearly full of water with a coin at the bottom, that you have to cover with another coin.' 'And there's "Find the Whole Egg" in a tray of sand, and all but one of the egg shells are broken. And there's rolling pennies—only actually it's tuppences. And there's two kinds of darts, and Ducking for Apples. And there's the toy stall, and a book stall—they've got lots of annuals, Beezer Annuals and Brownie Annuals and all sorts.'

'Oh,' I said, and went to look through the books; and all too soon there were banging noises as the stalls were dismantled with some vigour and taken into the canteen to serve their usual purpose as dinner tables.

Tuesday 1 *July*

The villagers could find nothing good to say about the weather. The wind and biting cold were too much for their spirits. 'Not a bit nice, is it?' they said; and the more sophisticated said, 'Isn't it *unspeakable*?'

The children seemed unaffected. A few of the girls wore gabardine raincoats or zipped jackets, but most had nothing extra on but a cardigan. The acrobats were apparently not deterred from performing in bare feet, although feet and legs were mottled blue with the cold. Crab-like creatures, midriffs exposed to the sky, scuttled to and fro. A girl was taking her shoes off and I watched her carefully to find out how one achieved the 'crab' position. She held her arms above her head and bent over backwards; she bent her legs till her body looked like the capital C in the clowns' alphabet, and then she was able to put the palms of her hands on the ground. Right way up again, she looked at her pitted palms and said, 'Most people do it on that patch over there, because it's smoother.'

Everyone seemed absorbed in what they were doing. I went to watch a boy who was scraping at the earth beneath the fence and might welcome an audience. 'I'm making a hide-out for our cars,' he said in his loud defiant voice. 'And I'm digging it out *with* my car—it's a James Bond car. Oh blimey, I've gone and scraped the 007 off it!' He fished in his pocket and brought out a pink-handled implement from a manicure set. 'Perhaps this'll be better,' he said.

The playground was noisier than usual. It was not only the excited and vociferous footballers; there was an undercurrent of sound from at least ten pairs of girls intoning clapping chants. I heard 'Under the bram bushes', 'When Susie was a baby', and 'When I was one I ate a bun'.[1] Then I noticed two little girls bobbing up and down neatly as they sang and clapped 'O Susianna'—a game I haven't heard here for about six years:

> I met my boyfriend walking down the road, O Susi-anna.
> And in his arms he was carrying a box, O Susi-anna.
> And in that box there was a dress, O Susi-anna.
> And on that dress there was a pocket, O Susi-anna.
> And in that pocket there was a note, O Susi-anna.
> And on that note there were four words, O Susi-anna.
> And those four words were 'Will you marry me?',
> O Susi-anna.
> And Susianna's reply was 'Yes I will', O Susi-anna.
> The next day they got marri-ed, O Susi-anna.
> And then they had a ba-a-by, O Susi-anna.

'And then we go very slow, and we say—

> And then her father di-i-ed, O Susi-anna.
> And then her mother di-i-ed—

'And we shout, "O Susi-anna!"' Each time they sang, 'O Susi-anna,' they touched their head, touched their shoulders, and bobbed down on 'anna'.

[1] For these three clapping games and the next, 'O Susianna', see I. and P. Opie, *The Singing Game*, 453–5, 458–61, 466, and 462.

When the bell rang the footballers found it difficult to simmer down. Three boys were sitting on a bench and the footballers clambered aboard shouting, 'Pile on, pile on, pile on!'[1] 'They're squashing people,' explained a more sober spectator. Soon the piling-on had turned into a fight, and the crowd were chanting, 'Da-*vid*, Da-*vid*,' as they backed their champion.

The excavator of car hide-outs was making himself felt in his own way. He was haranguing a group of girls who had wondered why Dennis was absent. 'Dennis is in 'ospital,' he yelled. 'Dennis had to have an injection. He's had an operation on his bottom.' 'On 'is chest,' corrected another boy. 'You're wrong, 'e's had a heart transplant.' A matronly girl standing beside me murmured, 'He talks as his belly guides him, that boy.'

Tuesday 8 *July*

Wet. The heavy downpour had only just let up, and the water stood in puddles.

The mouse boy and his two best friends came to meet me. The mouse boy is now intensely serious, though still only a third-year. 'We've made up a new club called The Saints,' he said, 'and we've all got nicknames. He is St Morphy, and he is St Sparky, and I am St Mouse, and we go round helping people. There's this iron called Morphy-Richards, so because his name is Richards we call him St Morphy. St Sparky—well, "sparky" means sort of quick, which he is. We've only got three members 'cos we don't want any double-agents. It's a sort of private detective group and we trust each other. We go round helping old people, and we help young people sometimes. Once we found a child in a supermarket who had lost his mother. We go out on missions on our bikes.' 'And we build tree houses,' said St Sparky.

Three little girls had been hanging around, waiting for the

[1] Cf. I. and P. Opie, *The Lore and Language of Schoolchildren*, 201.

mouse boy to finish. The one with corn-coloured hair and forget-me-not-blue eyes smiled at me winningly. 'Have you heard of the three holes in the ground?' she asked. She did not mind my knowing that the answer is 'Well, well, well.' Another charmer (hair tied on top of her head with a green bow) asked, 'Do you know what NESW stands for?' and said almost straight away, 'Naughty Elephants Squirt Water.' 'There's another for it,' said the first girl, 'Never Eat Shredded Wheat.'

Many boys were playing marbles (''Cos there's no football,' said one sourly), which left the football area free. Half the pitch was occupied by girls playing TV Programmes. The girl in front hummed and hawed over choosing a programme, and then called out, 'BP.' The girls standing in line could not hear, and there was much shouting to and fro before there was a cry of 'Blue Peter' and two girls raced to the wall and back, putting their hands and one foot against the wall before they turned.

In the corner by the steps some girls were running around waving their arms. They dissolved into giggles when I asked what game it was. 'It's not a proper game at all,' they said. 'It's only one we made up. It's called Noah's Ark. ['For some odd reason,' added one in a comically deep voice.] A person has to be on it, and they say, like, "roller skating". This place here is called the "Super Arena", and the people have to pretend to be roller skating and the person who's on has to see who's best at it. You can have two people on it, if you want.'

The sound of clapping games came from all around. Clapping and acrobatics must rate as this summer's big crazes.

Tuesday 15 *July*

St Swithin was uncertain what weather to allow us for the next forty days. The day started softly enough, with the clouds high, white, and ornamental. By playtime they had thickened and hung sulking overhead.

The senior boys clustered round the gate; some of them swung on it. 'We're waiting for the new first-years,' they said ogrishly. But when the new first-years came over, the older boys lost interest and gathered round me, instead, in preparation for a jokes session.

'Um,' they said, trying to raise the first joke. 'Um, who is it who's let down by his mates? A deep sea diver.' They are so tall now, I can look them straight in the face. One of the regular story-tellers of the past three years, now grown somewhat world-weary, volunteered a story about a French boy who came over to England to school. 'He went to London airport to see the planes take off, and he went to the zoo to see the zebras, and he went to hospital where somebody had a baby, and when he went back to school the teacher asked him, "Well, what have you been doing?" and he said, "Take off zebra, baby." Quite clever, isn't it?'

There was a lull. Out of the corner of my eye I could see some newcomers kicking a stone about.

'Oh, I knew there was another one,' said the boy who had started the session. 'What language do twins speak in Holland? Double Dutch. It's not frightfully good, I go in for Irish jokes mainly. There was this Irishman and he wanted to get over a river and he walked for miles and miles along a river bank and he saw this bridge and—he took it apart and made himself a raft.' He looked up and smiled. 'Yes, that's good,' I said, 'but I suppose I'd better go and see what the new first-years are doing. Thanks.'

A *very* small girl who could only be a visiting infant was in the grip of two natives and was yelling, 'Help me, Karen!' as loud as she could. She was literally saved by the bell. As if that was not a bad enough experience I thoughtlessly asked her what she had been playing this morning. She looked towards the gate and said in a choked voice, 'Line up!' 'Oh dear, of course,' I said, and we went quickly over to where the new children were silently lining up.

No teacher had yet arrived, so I asked the less frightened-looking ones what they had been playing. A tiny girl said, 'Orange Balls,' and whispered that 'this girl' and 'this girl' and 'this girl' had played it with her. 'That's what you play in the Infant School, isn't it?' She nodded. A boy told me, 'We've been playing It. We tuck people.' Another boy, a robust little chap, said boldly, 'I was racing round trying to catch John

Lambert and we went smack into each other.' Most of them had a finger or thumb in their mouth.

A nice sprightly teacher came to take charge of them. 'You're all being jolly brave, aren't you?' she said, bouncing gently on her flat-heeled, rubber-soled shoes. 'Now—each time the bell goes, when you're at the Junior School, you line up on these two lines and be ready to cross the road together. Are you listening, you little ones there at the back? Now—we're going to take you and show you where the toilets are.' Another teacher arrived. The children about-turned and went marching off to see the toilets in the corner of the playground.

Index

W9-BNP-423

THE LAWS OF Invincible Leadership

THE LAWS OF Invincible Leadership

An Empowering Guide for Continuous and Lasting Success in Business and in Life

Ryuho Okawa

| IRH PRESS |

Originally published in Japan as
Joushou No Hou by IRH Press Co., Ltd., in January 2002.

IRH PRESS
New York

Library of Congress Cataloging-in-Publication Data

ISBN 13: 978-1-942125-30-3
ISBN 10: 1-942125-30-5

Printed in Canada

First Edition

Book Design: Jess Morphew

Contents

CHAPTER TWO 75

Developing Winning Strategies

Seven Keys to Continuous Success in Management and in Life

CHAPTER THREE 119

Cultivating a Management Mentality

Five Keys to the Development and Prosperity of Your Business

CHAPTER FOUR 149

Overcoming Recession

Six Useful Perspectives during Transition Periods

Preface

My philosophy of success and progress is an aspect of my thinking that distinguishes me as a religious leader, and it is to my great delight that I've had this opportunity to compile my theory of success into a single volume in this book.

Many religious leaders teach about happiness in the other world, and I am one among them. But very few have been able to offer complete teachings about living happily in the present life. Their principles often turn out to be short-lived because whether their principles work or not soon becomes obvious to everyone. So I have consistently talked about happiness in this world that we can take with us into the next world. The Buddhist belief that our state of mind in this life determines where we must go in the next life is at the basis of this idea.

I want to tell all the people of this world about my wish for you all to find true happiness in this world and for your happiness to continue into your life in the next world. This is what my heart sincerely wishes. I have imbued the title word *invincible* with countless such hopes and wishes from my heart.

I pray with all my might that this book will serve to inspire millions of modern readers to have courage and that it will cast the light of wisdom upon countless generations to come.

Ryuho Okawa
Founder and CEO
Happy Science Group

CHAPTER ONE

Building a Strong Foundation for a Lifetime of Success

Seven Secrets *to* Becoming *an* Effective Leader

◀ ONE ▶

Set Virtuous Aspirations

AIM FOR SUCCESS that Brings Happiness

The universe offers an immeasurable array of guiding principles of success. So in this chapter, I would like to introduce seven essential secrets to becoming an effective leader, particularly for young professionals as well as others who aim to become successful leaders at work.

These days, the words *success* and *happiness* are often used synonymously. But as we look into people's real lives, we may find that success and happiness can mean different things to different people. For example, some people in the world of business are willing to gain advancement in their career lives at great cost to their families' happiness, while there are also others who willingly give up prospects for career advancement to protect their family's happiness or their own inner fulfillment.

When we consider these things, it could seem odd to say that happiness and success go hand in hand. But many people still regard them as synonymous ideals, and I also agree that there are some aspects of success that can be unified with happiness.

As the founder of Happy Science, I developed my philosophy of success to offer guiding principles that can help us merge success with true happiness to as wide an extent as possible. The following secrets of success are based on my belief that happiness is not just a result or outcome we attain when we reach the pinnacle of success; happiness can be also experienced constantly as we advance along the path of success. In the same way, our journey to success should also let us savor the inner fulfillment of happiness, and this is the kind of success I constantly hope will be fostered and grown throughout our societies and the world.

ASPIRE TO SPREAD HAPPINESS through Your Leadership

According to a commonly accepted method of success, we need to make many sacrifices to climb to the highest heights of success. So, there are many people who set goals of advancing to high positions of power and status so they can achieve successful careers as policymakers, government ministers, presidents or prime ministers of nations, head executives of corporate institutions, artistic geniuses, and many other types of vocations.

In many of these cases, the rise to such positions requires much hard work and tearful sacrifices made by a countless number of people. If the efforts and sacrifices of these supporters are requited in some way, then their sacrifices will have been well worthwhile the hardship. But the reality is that they sometimes receive little repayment for what they contributed, even when the people they worked for become successful.

As we aspire to become successful in our own careers and goals, we do not want our successes to be the result of many people's unrequited sacrifices. Rather, we want to see that our success leads to other people's successes and our happiness leads to many other people's happiness.

Then first and foremost, we need to ask ourselves, for what purpose are we striving to reach our goals—what are the underlying aspirations or mindsets of our hearts? This is the most important step of attaining the kind of success that brings true happiness. If we find that our wish to succeed lacks consideration for others' welfare so much that we don't mind treading over others and making them unhappy as long as we attain wealth and influence of our own, then this mental attitude will lead us away from the ideal shape of success.

There are many books on this worldly kind of success that promote the use of aggressive methods, which can often be an effective way of reaching our goals. For example, by applying aggressive sales and marketing tactics, we can take a failing product unwanted by customers and still achieve top sales results. Doing so can lead to a stellar performance, which can help us gain the favorable recognition of our superiors and eventually a promotion at work.

But, as I said earlier, this idea of success goes against the ideal kind of success that I encourage everyone to aspire to. Rather, our advancement to higher positions of power and a greater state of wealth should produce beneficial value to those around us, to our societies, or, on a larger scale, to the growth and prosperity of national and international economies around the world.

When our only aspiration is to be promoted to higher places, we start caring too much about avoiding making mistakes. The goal becomes making sure that we don't fail while we're in our position, even if this can potentially lead to problems after we leave our post. We stop caring that our neglectful actions could eventually come to light at some point in the future and put people in difficult or problematic situations, and we do this because all we're thinking about is finding ways to escape from making mistakes now.

So, let's say that you are promoted to an executive position in a manufacturing company you are working for, and you face the temptation to cut costs on pollution prevention, as this will help increase the company's overall revenue. When we face such a situation, we need to resist making the wrong decision of cutting corners for the sake of business profit, even if we believe that no one will notice. It is the heart and intention that we put into our decisions that is crucial, especially when we work in positions of responsibility.

It is the heart and intention that we put into our decisions that is crucial, especially when we work in positions of responsibility.

The same principle of aspiration holds true for those who hold public offices. Public officials are often guilty of valuing their career longevity and their own career advancement over being honest about their mistakes. They are reputed to conceal their mistakes, gloss over issues that have gone wrong, and avoid getting involved with policies that carry a risk of failure. In their hope to preserve their public reputation, they settle on inaction and deferment as their method of escape.

In my home country, Japan, high-ranking officials are often instructed on their first day in office to postpone issues that could be handled the next day. They begin their first day as public servants deferring decision-making altogether so as to avoid making mistakes and getting involved in complicated issues. And they continue to do so throughout their terms in office. As a result, the issues at hand that need to get resolved do not get addressed properly, and this endangers the well-being of the citizens who are ultimately the ones who suffer the most because of their officials' inaction and indecision.

Seeking success for personal protection is like wanting the head executive position in a company solely for the power and wealth that comes with it, and this mindset should raise concern. It is true, of course, that we are promoted to higher positions of responsibility and recompensed with greater wealth as a natural reward for our achievements. But though these rewards serve as report cards acknowledging the quality of the work we have accomplished so far, the "A" grade we earned does not improve our intelligence to help us perform better in the next semester. Rather, the "A" is the result of our efforts from the previous marking period or semester.

In any position you hold in any career you choose, the premise for your desire to succeed should be an aspiration to fulfill as many people's happiness as possible and work for the sake of other people's benefit. But if achieving your aspiration might result in spreading unhappiness and forcing sacrifice on many people, heaven will not want to bless you in your goals, and you will need to either change your aspiration or stop seeking promotion. It is my hope that an increasing number of people who achieve success will live and work with virtuous aspirations of success.

LIVE YOUR LIFE
with Positive Purpose

To become a successful leader, you need to set an aspiration for success in this world that does not bring harm to those around you. Your aspiration should be based on a desire to make your time in this world as meaningful as possible as you use your abilities and God-given talents to the fullest possible degree while, at the same time, contributing to transforming the societies, countries, and world around you, no matter how modestly, into ideal societies for all.

Though we may not have the power to change the entire world in an instant, we certainly are capable of contributing a bucketful of benevolent work to humanity's river of ideals.

This aspiration may sound old-fashioned and unpopular in these modern times; it may remind us of the morals we learned in history lessons and the ancient fables that we read long ago in childhood. This idea of success may seem out of sync with the pleasure-seeking times that we live in today. But the spiritual laws governing humankind unmistakably exist, and this is a truth we will understand as plain as day on the day we perish from this world.

And when we do, we'll also see clearly that our lives fall into one of two basic groups: lives lived with positive purpose and lives lived with negative purpose. Which of these groups we fall into is determined by which of two forms of success we aspire to achieve. If we choose to live with a harmful aspiration, the authority, wealth, or status that we achieve will mean that we have committed greater misdeeds due to the greater extent of our influence, and therefore we'll ultimately reap graver consequences for ourselves. At that point, however, we'll no longer have the choice to redo our lives in this world.

There are many books that encourage us to realize our goals even at the cost of taking others down as an essential method that leads to success. But these approaches are often inspired by ideas focused on the self, and they tend to lack a heart of love for others. They may be effective principles for achieving self-oriented forms of success, but this is not the ideal or true shape of success we want to realize in this world. This is why having the right aspiration is a key principle of success and essential to building a strong foundation for a successful life of leadership.

◀ TWO ▶

Develop a Life Strategy and a Set of Tactics

UNDERSTANDING THE DIFFERENCE between a Strategy and Tactics

The second secret to becoming an effective leader is to outline a strategic plan for your life and then devise a set of methods for fulfilling it. A strategy, in general, is the broad aims and goals we want to accomplish in the future. So a life strategy is a master plan that serves as our essential guide through our lives. The methods or tactics, on the other hand, are shorter-term ideas that determine specifically what needs to be done and how to fulfill our strategy. To help distinguish between them, you might think of your strategy as the conceptual element of your plan and the tactics as the tangible elements that you can see happening as you are carrying them out.

Finding the distinction between a strategy and tactics may be hard to do, and this, in fact, is often the reason why highly talented people find themselves not succeeding. They usually confuse the methods with the strategy. They mistakenly prioritize working on the specific techniques and tactics in front of them over working strategically toward winning the long-range victory. An incisive aptitude allows them to pay close attention to technicalities and details, but this often leads to focusing on smaller issues and neglecting to keep their sights firmly set on the bigger picture. Their thinking pattern resembles technical experts or those who work in very specialized jobs.

This trait could make it difficult for them to succeed when they work in positions above others. But working with a boss who can put their skills to good use may allow them to find the path to their success.

So the ability to tell the strategy apart from the methods is essential to success, and this ability can be cultivated through effort. To better understand the distinction between developing a strategy and devising the methods, let's consider a basic example of a girl in elementary school who wants to grow up to be a doctor.

If she is to achieve her dream of becoming a medical professional, going to medical school will be an essential strategic goal. So as part of her strategy, she will need to determine her principal choice of university and set earning admission to that university as her primary goal. To decide which university she wants to attend, she will need to begin by looking at a range of undergraduate universities offering degrees in premedical education or other qualifying majors and determining the ones that best suit her or her parents' financial capacity.

If she or her parents are able to afford a higher rate of tuition, she can include many private institutions in her list of options. But if this is difficult to do, her choices will mainly come from state universities and community colleges that offer premed courses.

If a state university works best for her and her parents, then the final deciding factor will be based mostly on where she currently lives.

And, after she has narrowed down her choices of universities in this way, she will move on to the next step of looking into which middle schools and high schools will best prepare her to gain admission into the university she has chosen.

In addition to determining the educational qualifications, another essential aspect of outlining a successful strategy will be to make sure that a career in medicine really suits her as a person. For example, the ability to do well in school is not the most reliable indicator of whether someone is compatible with a medical profession. Sometimes, students who decide on pursuing a medical career based solely on scholastic achievement find that they are unable to handle their first autopsy at medical school and pass out during the middle of the operation.

Another aspect of career compatibility is that doctors spend all day with patients. So finding a sense of fulfillment through constant contact with people is another quality she should think of as essential to this career. If she feels uncomfortable about being around and interacting with people all the time, she may want to reconsider her choice of career.

As my description so far indicates, gaining an understanding of and assessing her compatibility with the day-to-day aspects of her dream profession are key to devising a career strategy. Her strategy will need to be based on this assessment of her aptitude for becoming a doctor.

If she determines that she still wants to become a doctor, she is now better prepared to devise a successful strategy. So if we return to her educational strategy that I began discussing earlier, she will need to outline a plan that will successfully get her into medical school. This can be done by backtracking from her main choice of university to determine which high school and middle school will increase her chances of being admitted to each level of school above it. Some high schools offer curriculums designed for students pursuing premed programs in college, and these schools may be essential to her chosen university. Then, she could look into which middle school she'll need to attend in order to be accepted by her chosen high school.

In addition to this, she should determine her main academic goals for her grade school education. For example, she will need to focus a large part of her studies on performing well in math and science. As she outlines her general goals leading up to her dream career in this way, her overall life strategy will start taking shape.

The next step she will want to take is to determine her methods or tactics. Her methods are the detailed action plans that she will choose and alter as the need arises to increase her chances of success. This is what differentiates the methods from the strategy.

For example, how well she needs to do on her next exam and what her method of preparation will be, how she'll go about improving in her weak subjects, and how much improvement she needs to make in her strong subjects are all decisions about tactics. Hiring a tutor, for instance, is an example of a tactic she can consider.

In conclusion, the strategy is the plan that helps you attain the long-range purpose, and the methods are the specific tactics and techniques you will implement to help you successfully reach your goals.

DEVELOP A STRATEGY
for Your Career

Practicing strategic thinking may seem as if you are just planning for a vague, intangible dream in the distant future. But this dream is certain to unfold into reality as time passes. Strategic planning is as essential to our own success in life as it is for the girl in this example. So we also need to spend time planning the course we would like our lives to follow and determine what we should be doing now to fulfill them.

The first important stage of strategic planning in your own life involves deciding on the career or vocation you want to work in throughout your life. This, then, becomes the basis for envisioning the kind of family and status in society you want to build in the future.

Here, I would like to point out once again that as you develop your life strategy in this way, you should also remember to ascertain your life's ultimate purpose. As I mentioned in the beginning of this chapter, this shouldn't include ignoble aspirations that give rise to other people's unhappiness.

Your life strategy should serve as your master plan over several years to several decades. Since a strategy defines the main aim that will guide you through your life, you do not want to have to change it very often once you've determined on it. So, in other words, there is no need for your strategy to be planned out to the last detail; instead, it should be a vague, general image of the next several years or decades of your life.

How far into the future do you want to plan for? Your plan can certainly span your entire life, which would be about fifty to eighty years. To live a successful life, we each need to have some overall vision of how we want our life to appear as a whole, with a general image of what we want to have accomplished in each decade of life.

For example, you may have specific career goals that you want to achieve by the age of thirty, in terms of your career accomplishments and further education. You also should consider whether you hope to get married or stay single throughout your life. If you decide to include marriage and building a family into your vision of the future, your life strategy will want to take the number of children into consideration.

Then, apart from deciding the kind of upbringing you wish to give your children, you may also want to determine how much time and money will be required to raise them and send them to college. There are other additional factors to consider, such as the type of university you want to send them to and the kind of vocation you want them to pursue.

Taking each decade and setting general goals for each one allows you to think strategically from a broad perspective about the purpose you want to fulfill in this world.

Taking each decade and setting general goals for each one allows you to think strategically from a broad perspective about how you want to live your life and the purpose you want to fulfill in this world. This is a vital aspect of living a successful life.

LIFE IS A CHAIN
of Decisions

After you've developed your general life strategy, you must also define the specific set of methods and tactics you'll use to bring it to fruition.

The methods can take any number of various forms, so I recommend choosing the ones that appeal the most to you. They can be changed as often as you need them to, as long as they contribute to the main strategy and leave it essentially intact.

The most important thing you need to be aware of is that the tactics exist to serve the success of your strategy. This conscious awareness will strengthen your resolve about the life strategy you set in place, allowing it to take firm root inside your will. If you maintain this awareness, you won't let small issues shake your determination; your mind will remain fixed firmly on the grand vision you had set in your master strategy as you gain small victories on the path forward to your life's success.

The way your life strategy is constructed and the course it follows closely resemble the workings behind the binary system in computer technology. Just as computer technology uses strings of zeros and ones to express the information that's put on the screen, our lives are constructed by the chain of small decisions we make by responding with a yes or a no to each of life's abundant topics.

Every small decision is strung together with all our other decisions to create a unique chain of successes and mistakes. On a daily and yearly basis, we make decisions in response to each of life's events; we choose to affirm or deny what arises in front of us, decide to look at them in a positive light or a negative light, and determine on this option or the other option. As these decisions accumulate, they build upon one another to shape the unique lives that unfold before us.

So, our ability to make the right decision about each problem we are faced with is what determines the eventual outcome of our lives. Sometimes we make the wrong decision, sometimes the right decision, and sometimes a string of right decisions. Our lives are the result of the unique combination of these decisions we make in each precious moment of this life.

◀ THREE ▶

Receive Spiritual Guidance

MAINTAIN a Pure and Open Heart

In the previous section, I explained that a strategic perspective on life is an essential element of success. Planning your life strategically will allow you to define your main goals and objectives and devise a clear plan for accomplishing them. But the ability you develop in this world alone is often not enough to reach these goals. So I would like to introduce the third secret to achieving success, which is to receive spiritual guidance from your guardian spirit and guiding spirits of heaven. Since the invisible world has been recognized to exist and to be exerting its spiritual influence upon this world, it would be a waste not to let this beneficial power assist your journey to success.

A guardian spirit is a benevolent spiritual existence, a brotherly or sisterly spirit of your own soul who accompanies you through your life to give you spiritual guidance and protection, and we each have one, without exception. And in addition to your guardian spirit, there are also guiding spirits in heaven who possess even higher spiritual capacity and wisdom and can offer further guidance. They are capable of providing advice when you take on a position of significant influence or of a high degree of specialization of work. They are leading experts in their specific fields and are highly qualified to lead those on earth to spectacular achievements. They exist in heaven in multitudes, and receiving their support is truly indispensable to finding your way to great success.

Then what is the key to receiving the spiritual guidance of your guardian and guiding spirits in heaven? To receive their guidance, the key is to remain pure in heart and open-hearted. You need to abandon self-focused desires and humbly wish to serve the power of God, angels, and the divinities of heaven so their ideals can be realized upon this world. When you align your heart with theirs, the magnitude and influence of their guidance upon your day-to-day life become multiplied. And as you continue making the effort to achieve this state of mind, you will develop a spiritual sense that will allow you to recognize when you are receiving their divine support.

> When you align your heart with spirits in heaven, the magnitude and influence of their guidance upon your day-to-day life become multiplied.

On the other hand, if you adopt an egoistic way of life, allowing your heart to become clouded by negative and impure thoughts and desires, this will gradually force your guardian and guiding spirits to distance themselves from you. You have probably encountered people leading egoistic lives who made you want to avoid them. The same thing happens with your guardian and guiding spirits when your heart becomes too egoistic. They are naturally sensitive to the quality of the vibrations you emanate and have difficulty associating with unholy inner vibrations. They will want to keep their distance from you as much as possible if your inner vibrations are not in tune with theirs, and, out of exasperation, they may even leave you to do as you wish.

If you allow this to happen, the danger is that you could invite the negative influence of hostile or malevolent spirits into your life. Those who go about their lives under the influence of malevolent spirits are those who are commonly regarded in society as vicious people. And they are capable of achieving a certain degree of material success in this world. But, as I discussed earlier, though you can reach a certain level of success by directing a strong will toward a set aim and employing aggressive methods, you need to take caution not to follow this path to purely worldly success. The high recognition and acclaim you earn on this path seldom last into the other world. A life lived under the influence of vicious spirits often invites a bad reputation, a sense of isolation, aloneness, anxiousness, and others' dislike, immersing your twilight years of life in much unhappiness.

Intelligence and cleverness can also bring us success, and since they are often independent of the good or evil in us, they can help us succeed even while we are under evil spiritual influence. But this path to success is not the sort of life you should ever seek. For example, aspiring to succeed through professional thievery can attract many spirits with past lives involved in crookery. This occurs because the spiritual vibration of your aspiration invites spirits with like vibrations.

In contrast, you are capable of receiving the spiritual guidance of your guardian and guiding spirits even while you are pursuing grand ambitions of success if you are willing to put in hard and honest work, are devoid of self-willed desires, and want to humbly accept their spiritual guidance.

And when your efforts open up avenues to a bright future, always remember your gratitude to them, though they may remain invisible to your eyes. When your guardian and guiding spirits receive your deep gratitude, they will feel glad to have supported you and will wish to give you the further assistance you'll need to keep succeeding.

GROWING OUR SOULS through Successes

The way our minds work is truly remarkable. Simply thinking of someone is all that we need to do to spiritually connect ourselves to them, and it only takes a second. Physical distance makes no difference in the time it takes our thoughts to reach them. Just by the simple act of thinking, we create a spiritual link just as if we were placing a phone call.

By not keeping your guardian spirit in your thoughts, you leave yourself empty of the means to communicate with them. On the other hand, you will always have a secure channel of communication as long as you continue to hold them in your thoughts. As long as you continue to be aware of them in your heart, your guardian spirit will always know your mind.

I truly believe that all of us will benefit greatly from receiving the spiritual support of our guardian and guiding spirits. There are many benefits to us and them when we live our lives under their divine guidance and support.

From the perspective of our guardian and guiding spirits, by supporting those of us who are living virtuously and in purity in this world, they're allowed to expand and enhance the breadth of their own virtuous qualities of character. Helping us create an ideal world on earth holds deep spiritual value to them. It allows them to increase their own benevolent power, and they feel appreciative when they have an opportunity to do so.

Second, this also lets them gain valuable experience to learn from. Guiding an individual on earth to accomplish something of true value to society that flowers into unexpected proportions of success not only lets them gain experiences that bring fulfillment to themselves, but also contributes to the work they do in heaven.

And from our own perspective, the success we achieve as we live under divine guidance lets us widen the capacity of our own souls. Our achievements allow us to build our confidence. And this is the valuable kind of self-confidence that we can bring with us to the other world. Not only that, but it will also serve as a source of precious strength in our next incarnation in this world.

When we live a life of many failures, we inevitably fall on the receiving end of others' aid. But by living a life of many successes, we empower ourselves to be on the side of giving and helping others grow. This is one way we cultivate leadership, and it is also what makes being successful wonderful.

◂ FOUR ▸

Get Support from Others

ENHANCE YOUR SUCCESS through Others' Support

The fourth secret to becoming a successful leader is inspired by a military analogy: success requires the skillful maneuvering of troops. It often is the case in this physical world that even when we hold a strong will to succeed, success can be hard to come by when we depend on ourselves for everything.

Those among the successful may feel that their own hard work was the reason for their achievements. And there are also others who have jobs that require working alone. But even in these cases, their successes were achieved with the support of many other people, such as their family, friends, neighbors, and others who willingly lent them support.

It is seldom the case that we achieve success all on our own. Let's look at an example of an accomplished musician. It's true that he practices and performs alone on his instrument, so he may attribute his own exceptional talents and efforts to his rise to success and acclaim. But, in reality, an accomplishment like this requires the hard work and dedication of many other people who help make dreams like his come true.

We can find the same principle of support underlying jobs we find elsewhere in society. The most common type of job we find today is in a company or other type of organization where we are required to work not alone but with a group of many other people. Although these jobs often put us in charge of certain aspects of work that we carry out on an individual basis, from a broader perspective, each job functions interdependently throughout the company and contributes to running the daily, organic operations of the whole.

This means that when one person performs poorly, other workers are affected, and someone will need to make up for that poor performance. In the same vein, the work performed by highly effective workers positively affects their surroundings and ultimately contributes to the smooth operation of the company as a whole.

It's rare that we find an isolated, self-sufficient job anywhere. Often, we are falsely imagining in our mind that we are working alone toward our goals.

To succeed in reaching any sort of aim in life, we need to garner the help of a multitude of friends and supporters. Just as the wisdom of divine guidance is an essential component of our success, we also need to gain the help of supporters in this physical world to bring our dreams and goals into reality. Usually, this assistance comes from our superiors, colleagues, and subordinates at work or our spouses and children at home. Their support becomes indispensable as we work to bring great undertakings to life. Often, their support not only gives us the additional strength we need to succeed, but also lets us *multiply* our strength by twofold, threefold, tenfold, or even a hundred-fold to ten thousand-fold.

This is not at all an exaggeration. When we advance to an important position such as the chief executive of a company, we could have as many as ten thousand employees under us to position and direct strategically, as we do in a game of chess. By pointing everyone in a set direction, we are able to achieve far more than what we would have accomplished singlehandedly. This means that whether we achieve success depends to a great extent on how well we position and maneuver the people who work for us.

DEVELOP AN EYE for Outstanding People

In the context of a nation at war, appointing the right person to fill a crucial leadership position, such as that of the commander in chief of the combined naval fleet, can lead the nation to a decisive victory. But the failure to do so can lead to the nation's defeat and a half century to a full century of hardship. In this respect, when we examine how the Japanese military chose its leaders in World War II—its generals, lieutenant generals, and major generals—we find that there was a major flaw. The mistake was failing to choose the right men to fight in real live war; instead, they appointed men to leadership positions based on a bureaucratic system of promotion.

In contrast, Japanese military leaders during the Russo-Japanese War did something different. They commissioned Heihachiro Togo, someone who hadn't been considered elite material until then, to be the commander in chief of the combined naval fleet, bypassing other high-ranking men above him. Togo was recognized for having luck on his side, demonstrating virtuousness of character, and faithfully carrying out the orders he received from the headquarters, and these became the reasons he was singled out among other men to be promoted to this position. Until then, he had been considered too small in frame for command and had demonstrated little sign of military brilliance. So by the time he was appointed to be a colonel, everyone expected him to be heading toward retirement in the near future.

But, as history has shown, he was a man fit to be a leader. The decision to appoint him commander in chief of the naval fleet became the decision that led Japan to victory in the Russo-Japanese War. If the Japanese had chosen someone else to be their general, they may have well suffered a devastating defeat against the Russians.

We need to cultivate the ability to recognize the talent, ability, and good fortune that lie deep within people's souls.

It's often not enough to rely on our abilities alone to be successful. A vital element of success is found in choosing the right people to support us.

To find this kind of support, we need to develop an eye for noticing outstanding people. This means that we need to cultivate the ability to recognize the talent, ability, and good fortune that lie deep within people's souls. Credentials and personal history may indicate what the individual accomplished in the past, but we cannot determine other aspects, such as the unique auras they emanate, by looking only at their background. In other words, we also need to discern whether they are receiving the support of their guardian and guiding spirits in heaven.

Those who are capable of receiving divine assistance gain a variety of spiritual influences to help them accomplish more than they can alone. And placing such people into positions of leadership attracts many further contributors who want to offer their support.

But when someone under the influence of malevolent spirits, not divine spirits, is appointed to a leadership position, it will lead to the opposite result. This person will tend to attract others holding malicious vibrations to aid his endeavors.

In this sense, the ability to see the inner qualities of the people we enlist is vital to our success. In addition, we gain a tremendous source of strength when the people we appoint to leadership roles share in our aspirations.

DELEGATE
Work

Most of us in this modern age have jobs that belong to a company or other kind of organization. So assuming that many of my readers work within corporate institutions, a key to advancing to a leadership role or becoming a successful leader is your ability to delegate your work.

There is only so much that we can accomplish on our own. No matter how outstanding we are at what we do, it can't compare to what we can accomplish by employing a hundred more people, or even just ten more people. This is why effective delegation of work is a crucial component of leadership success.

A key element in effective delegation of work is understanding the skill sets and qualities of the people you are delegating to. This understanding helps you determine which jobs are best suited to them and which ones are still difficult for them to do. And when you begin, one thing you will need to be aware of is that it can seem unproductive in the beginning, because not everyone is capable of the quality you produce yourself. But this is a reality that you'll need to accept as a necessary aspect of successful leadership.

Even if those who work under you can deliver only 70 to 80 percent of the quality you produce yourself, this drawback is outweighed by the larger results that they help create for the company as a whole, especially when you compare that to what you'd be accomplishing all on your own. In delegating work effectively, what's important beyond anything else is the valuable time it frees up for work of higher value requiring advanced skills.

The same principle can be applied effectively in religious organizations, too. For example, Robert Schuller was a Christian evangelist from California who achieved enough success to build a church made out of glass known as the Crystal Cathedral. What he said about how he accomplished such success for his church is very interesting, considering that it's coming from a religious leader. He said that the ultimate key to growing your enterprise is hiring smart and capable people to handle jobs that you don't have to do yourself. He said that this is possible if you have the money to hire good people. As long as you can find the money, you can hire smart people to work on the practical aspects of the enterprise while you devote yourself to jobs that only you can do.

What was the job that he considered only his? To him, this job was giving Sunday sermons. This was one job that he knew he couldn't delegate to others. He considered the work that he devoted every Saturday and Sunday to—on Saturday, choosing a topic to talk about and creating a draft of his sermon, and on Sunday, delivering a worthwhile weekend service for his congregation—was a job only he could do, and not something he could delegate to others to do on his behalf.

The Crystal Cathedral was an enormous enterprise that must have required as much as $20 million to build. An endeavor of this scale requires running multiple business projects simultaneously, including fundraising, architectural design, and construction, followed by maintenance and management. These projects all require heavily administrative and practical work. It would have been impossible for Robert Schuller to oversee all these projects on his own and still have time for spiritual study and to perform his main job of spiritual leader effectively.

This was how he came up with the idea to hire smart and capable people to manage these projects for him, and by doing this, he created time to concentrate on the high-value job that only he could do. He knew that this could be done as long as he had the money to offer his employees appropriate compensation.

Robert Schuller's idea worked, and he and his church achieved a certain degree of success. Most churches are not profiting the way he was able to simply because the fundraising, maintenance, and sermons are all managed by the pastor himself.

Delegation of work is as vital to the management of religious organizations as it is to business enterprises. Regardless of the nature or content of their mission, huge enterprises of any kind can succeed by hiring capable people to do good work. As your employees succeed at what they do, so will you enlarge the scope of your own achievements and successes.

DEVOTE YOUR TIME
to Jobs of Higher Value

If you are struggling because your capabilities at work have hit a ceiling, then you may require a different kind of solution. But if this is not the case and you're capable of further growth and success, then your solution is to delegate responsibility. You need to entrust others who don't earn as much as you do with more of your responsibilities. This is necessary to do in order to create time for yourself that you can devote to more challenging, innovative, and advanced projects and ideas.

Your workload will only keep increasing as time passes. So if you find that you are feeling overwhelmed at having too much on your plate, then your work most likely needs to be reviewed and restructured. Making a thorough review of your work should reveal many tasks that can actually be handled well by others around you.

When you have done this, the next step is to choose someone who is capable of taking on each of these tasks. Even if you feel unsure of the person's ability to handle a task, you can still let them have a try. If, as a result, they do find that the job is not suited to them, then you can shift course and find someone else to delegate the work to.

Delegating responsibility frees up a remarkable amount of time. The purpose of creating free time is not to use it to do nothing, of course. The purpose is to search for new emerging opportunities, tackle difficult issues, and explore new ideas and business markets holding profitable potential for your company's future. This will lead not only to your own success, but also your subordinates' growth into more advanced levels of work.

If you cannot seem to move up in your work even though you are performing well, the reason is most likely your unwillingness to delegate responsibility. When we become too attached to our work and can't let go of control, our companies reach their limits when we reach the limits of our own capabilities. This leaves little room for other employees to take on more effective, productive responsibilities. We end up working for the salaries we ourselves are earning, while others are left with nothing productive to do. This tendency can be found in capable people more often than we think, and they should take caution against this tendency as they work their way up the ladder of success, because they are bound to face this situation somewhere along their path to success.

◀ FIVE ▶

Be Quick to Seize Opportunities

BE DECISIVE
When Opportunity Knocks

The fifth secret I would like to discuss is to quickly seize opportunities. We're constantly presented with choices that can lead us to either success or failure. But those who are currently riding the waves of success are those who were quick to seize upon opportunities when they emerged.

We can never truly tell when an opportunity could show up in front of us. We can spend ages waiting patiently for an opportunity to arrive, only to find not even the slightest approach of one. But much like the waves of the ocean, it's often the case that when an opportunity finally comes to us, we'll find many others appearing in rapid succession. If we show any hint of indecisiveness in these moments, we may let these opportunities escape us forever.

It's just as they say: we need to grasp hold of the goddess of success by her bangs, because we'll find nothing on her head to grasp hold of from behind her. It may seem strange to imagine a goddess who only has bangs, but, nevertheless, this is a telling image of success's ephemeral nature. Opportunities for success come and go so quickly that we'll miss them if we don't recognize them quickly. It's all too common to find people who spent many years in regret for not having recognized when opportunity was knocking on their door.

When it comes to the ability to seize opportunities, everyone generally falls into one of three levels. The advanced group consists of those who are perceptive enough that they can identify an opportunity as it is approaching them. The intermediate group is composed of average people who notice an opportunity when it appears in front of them. And the beginner group recognizes that something was an opportunity only much later, three to ten years after it has vanished.

Those in the advanced group have a knack for seizing on almost every opportunity that comes up, but those who fall into the third group have a tendency to consistently let opportunities get away. What do we find in common among those who frequently let opportunities slip past them? The simple answer is that they are not decisive enough. But I can say that the people in this group resemble someone who wants to stay inside a lukewarm bath. They are afraid of getting out and risking the coldness, even while knowing that they could equally catch a cold by remaining there in tepid water.

ABANDON THE PAST and Take Hold of the Future

Many of those who miss a lot of opportunities for success do so because they are afraid to let go of past successes, and this doesn't allow them to make new decisions. There are times when we have to let go of the past to take hold of the future. If we keep a firm grip on what we have now, our hands will be too full to take hold of new opportunities. To be successful, sometimes we have to shift our minds away from what we want to keep or achieve and decide what to let go of. This is an essential aspect of achieving success that we must learn to accept.

> Many of those who miss a lot of opportunities for success do so because they are afraid to let go of past successes, and this doesn't allow them to make new decisions.

But this is often difficult to do, and that is because the decision to follow an opportunity often involves forfeiting other possibilities. When you decide on the vocation you want to pursue, you are also forsaking the possibility of working in other professions, at least for a while.

The same holds true when you decide to get married. Making this decision means that you have chosen someone to be your partner for life. It means you are prepared to sacrifice possibilities with all other prospective partners and set firm roots down in the commitment to be with one person.

These choices may feel scary to some people. But they are necessary decisions. By prolonging your attachment to other men or women you have had relationships with, you could be letting the chance of a very happy marriage escape you. Those who frequently miss opportunities for success tend to do so because they are holding on too tightly to what they already have and are not thinking of what to let go of.

Another common example that requires the courage to let go can be found in relationships between parents and children. The strong relationship that develops between an only child and her parents makes it difficult for some people nowadays to leave the nest, build financial independence, and start a family of their own.

Only children may feel as if they are being ungrateful by deciding to leave their parents' sides, and they may also have misgivings about dealing with conflicting values by getting involved in new relationships. But in this case also, they need to close the curtains on old relationships to be able to begin new chapters in their lives. When they grow old, they may regret not having taken this step and having left precious possibilities unexplored.

Giving up something old to gain something new is an important aspect of the Buddhist teaching of impermanence, but it is also a key component of modern principles of innovation. Innovation is about not only grabbing new opportunities, but also about systematically disposing of what we have now.

This is a vital principle of life to know about, because it is something we often need to do to begin relationships with new people. For example, the process of moving up from middle school to high school naturally ends our relationships with our middle school teachers. This is a natural part of the process of creating new relationships. When we begin working at a new job after graduation, we naturally have to part ways with our professors and friends in college.

Bringing our past to a close is a part of the natural course of starting a new stage in life. Starting life in a new place is much like a molting process. To begin life in a new world, we must shed our old skin and be prepared to part ways with our old friends and family.

The same principle holds true as we transition from adolescence into adulthood. We move on from our parents' guardianship and support and begin earning our own living, providing for our own meals, buying our own homes, and starting a family with someone new. Not having the courage to replace old relationships with new loved ones can result in pain from indecisiveness in the end.

So we need to be quick to seize opportunities when they arise. You don't want to let indecisiveness slow you down, because more often than not, this will lead you to make the wrong decision.

Instead, tune into the tides of change that are occurring around you. If you sense that a valuable career opportunity has come your way, you must grab hold of it. If you think that you have just met someone invaluable to your future, then you must grab hold of this person and begin building a relationship. Those who are slow to act when these opportunities appear will have difficulty achieving success, but all successful people have been quick to seize upon opportunities like these.

◀ SIX ▶

Develop an Unshakable Mind

GREAT ENDEAVORS Require More than a Sharp Mind

The sixth secret to leadership success is to cultivate an unshakable mind, a mindset of unswerving patience. Some fast-thinking people are very quick at seizing opportunities, and this allows them to achieve some level of success, but some of them may find that they eventually hit a ceiling. The reason for this could be that they possess diverse talents, interests, and abilities, but have only developed these to a shallow depth that does not lead them to true success. As college students and as employees, multitalented people demonstrate their gifted ability to accomplish many things, but that does not necessarily guarantee that they will reach the highest levels of the ladder of success.

For instance, in recent years, we have seen the rise of many global businesses in Japan, and English is now used everywhere, including in trading companies, banks, and manufacturing companies. But when we look closely at people in high positions, they are not necessarily fluent speakers of English. Instead, many people who are proficient in English are placed in specialized positions as translators for other people in the company and are often not promoted to higher positions. On the other hand, those who are not as fluent in English but have enough familiarity with the language to speak and read it sufficiently are able to focus on demonstrating their true capacity as leaders, and as a result, they advance to higher positions of management.

PITFALLS of Exceptionally Talented People

You can find a similar phenomenon occurring in other professions. For example, star journalists may find that they are not promoted very often to higher positions. Many reporters who went to the battlefields of the Vietnam War to report on breaking news developments became good promotion material for their employers but were not necessarily appointed to higher positions later on.

The recognition they earned came from having sharp instincts for sniffing out scoops and showing the courage to risk their lives, much as sprinters are recognized for their explosive speed. While these are skills that allow journalists to work well on their own, they are not skills that leaders of an organization would want to incorporate into the entire company culture.

A media organization won't usually send indispensable people to a dangerous place like the warzones of the Vietnam War where their lives could be easily harmed. It would much rather send people who won't bring great damage to the company even if they're lost.

People who are considered invaluable for the smooth running of the organization are usually kept at the headquarters and promoted to higher positions.

In our twenties and thirties, peers who shine out like outstanding journalists may appear as if they are outracing everyone. But it's possible for these roles to become reversed as we get older, in our forties and fifties, when those who originally came in second or third place begin to take on higher positions of management.

Ironically, those who display a strikingly exceptional talent are often those who have not developed their abilities deeply. And if these people rely heavily on these underdeveloped abilities to help them succeed, they may neglect to explore a wider breadth of potentials and abilities that could be developed for the future.

The ability to work effectively in management positions requires a balanced perspective. As you work with your team of many members, you need to be capable of keeping things in balance as you move forward and help everyone achieve their best as a team. Those who aim to manage people with this balanced perspective are those who are often recognized and appointed to higher positions of management in their company.

In contrast, those with exceptional talent and the ability to be effective on their own are often unable to entrust others with their responsibilities and unwilling to delegate work. They have difficulty taking on subordinates because they cannot let go of control, which ends up limiting the scope of their own work.

This is the reason why we don't see many star journalists training other junior reporters to mimic their method of success: star journalists are more likely to compete with younger reporters than train and educate them. When this happens, managers are likely to decide that it's more beneficial to the company as a whole and more productive for these reporters themselves to work in competition with one another. This is how people from a management perspective tend to use their star reporters. And this is the reason why exceptional reporters are usually not found in higher positions up the corporate ladder. This is something that many people should keep in mind in their own pursuit of success.

WITHSTAND THE TEMPTATION to Chase Only Small Victories

So, being quick-thinking, exceptionally talented, and decisive are essential attributes to cultivate, but if your goal is to become a central player in your organization, you also need to develop patience and perseverance.

Central players are like the unshakable trunk of a grand tree that withstands the passage of time. Central players don't easily lose sight of the larger victory they're aiming to win. Even if they win some and lose some at a tactical level, they always make sure that their broader strategy ultimately leads them to victory.

Getting wrapped up in tactical wins will not lead you to larger successes. For example, we should notice that the businessman who runs the chess club is working from a higher level of management than the chess player who is honing his skills to win every game. The owner may not be a good chess player, but he is practicing smarter business sense by operating a profitable business with a group of many chess players.

Getting caught up in winning the small games closely resembles the emotional thrill we experience from watching yo-yoing stock prices. Getting absorbed in this kind of emotional roller coaster can lower our chances of winning on a larger scale.

A sharp mind and the ability to work quickly are good traits to have in the earlier stages of our lives. But as we gain experience over the course of years, we should also learn to appreciate the value of the mindset of perseverance.

When we begin to take on larger projects, we will be faced with a variety of difficulties. We are bound to come up against opposing opinions, and we may not always win everyone's cooperation the way we would like to. When these situations arise, we should not give in to the emotional yo-yos we may feel tempted to spiral into. Instead, we should resolve to endure and be patient through times of difficulty.

It is crucial to our success that we develop the mindset to withstand and persevere against difficulties when they arise.

It is crucial to our success that we develop the mindset to withstand and persevere against such difficulties when they arise. By believing that patience upon patience will lead us to ultimate victory, we will discover that, over the course of one, two, three, or five years of perseverance, our inner strength has grown strong enough to help us actualize our goals.

PERSEVERE
Against Adversity

Then, how do we develop an unshakable mind? An unshakable mind is ultimately developed through effort. Working in society means that you'll be knocked about by many kinds of experiences. But your mind develops unshakable strength as you persevere and stand firm in spite of them, persisting until you achieve results. And the experience you gain by persevering in this way will lead you to do the same thing the next time adversity tests your determination.

The downfall with those who are quick-thinking is that they tend to seek immediate results, leading to a focus on smaller outcomes. They allow their minds to be shaken to and fro, and they gradually lose the strength to withstand the strain of these emotional swings. So you may find that the more of a fast thinker you are, the weaker your mind tends to be against the ups and downs of successes and failures.

As we gain experience over the years, there comes a time when we need to set our mental roots firmly into the soil and allow the trunks of our minds to thicken and grow taller. This may sound like I am encouraging you to do the opposite of being quick to seize opportunities. But cultivating an unshakable mind against the strongest gales and the most raging storms is also an essential key to success.

When we develop unshakable minds, we will find that many who suffer from chasing small victories will gather to us for shelter from the storm. They will depend on us to be staunch and steadfast leaders. So it's for their sake, too, that we avoid getting wrapped up in pursuing small successes. Instead, we must set our sights upon the ultimate victory ahead of us and focus on resolutely enduring the storm.

I can't emphasize enough that an unshakable mind is cultivated through the experiences we gain in this world. What is most vital is that when we prevail through one difficulty with an unshakable mind, this experience strengthens us to persevere through the next one. Those who are not as quick-witted, are slow to learn, lack foresight, and suffer from indecisiveness may ultimately find it too challenging to be successful. But if you are fast thinking but have a hard time reaching success, then what you need is nothing other than an unshakable mind. You should aim to become a staunch and unshakable tree that perseveres against life's many storms.

WAIT FOR TIME
to Be on Your Side

There is no escaping that life is not always going to be smooth sailing. It's important that we accept the constant cycle of good times and bad times that is inherent to human life. One moment we may be on top of the world, and the next moment we may fall to the depths of despair, and this is often how life was made to move forward. It is not at all uncommon to find that a huge rise to success is followed shortly by a downhill fall into failure. This constant cycle of ups and downs is a part of life that no one can escape from.

Some even say that life is made of biorhythms, such that life will sometimes bring us strings of successes and other times strings of failures. When we find ourselves in the midst of a train of failures, nothing seems to work out, no matter what we do. The best thing to do when we face these difficult times is stop struggling too hard to get out of the situation. We can gain victory when the timing is right and time is on our side. But in difficult times, we should expect that nothing will work out well, no matter what we do.

The secret is to wait patiently for the tide to turn and carry you forward again. This means that you first need to calmly consider whether this is the right time to play your cards. If you find that the tide of war is against you, it is the wrong time to wage a battle. Instead of needlessly dying in battle, this is the time to round up your troops, retreat from the lines, and wait patiently for time to take your side, for the next favorable opportunity to arrive.

There are times in life when there is nothing we can do to change the situation except wait a few years for the tides to turn. There are certain periods in life when we don't have to play our cards, and instead we need to use this time to prepare for the future by gathering and developing our strength.

This decision to retreat is certainly an enormously difficult choice to make. This is especially true when we are young; the more tender we are in age, the more strongly we feel the drive to keep on marching forward even in the face of death; and the more ground we lose in battle, the more strongly our passion to keep fighting burns on.

Nevertheless, if we know that the tide is against us, it is vital that we make the decision to retreat and wait for the next chance to arrive with an unshakable mind. This is the wisdom of an experienced general, which may be difficult for the young to accept at first. But developing this kind of patience and perspective is essential if you are ultimately to succeed in life in the long term.

As long as time is not yet on our side, success won't smile upon us, no matter how hard we fight to succeed. So we must absolutely wait under these circumstances for time to take its course.

If we succeed in waiting patiently for two or three years, this will prove that we were made for big achievements. This is difficult for most people to do, and they typically resort to doing something rash.

This reminds me of a simple lesson that we're taught to remember in swimming class in case we're ever in danger of drowning. When you are drowning, thrashing your arms in a panic will only lead you to swallow more water. Instead, you should calmly put your head under the water and allow yourself to relax and be still. Since your body is less dense than water, your body will eventually do its work to float to the surface on its own accord. Our bodies were practically created to float on their own, so it's actually difficult to drown when we're swimming in still water.

This is an important analogy for maintaining an unshakable mind when you are suffering through a cycle of failure. In a time of hardship, we must remember not to display a struggle and instead just think about becoming one with the water, allowing ourselves to wait to naturally float to the surface. This may be the skill of a seasoned expert at life. But it is possible to practice this by being aware in your heart that this is how life goes, and then you'll find that you're sailing more calmly across the currents of life than before.

In a time of such adversity, you may need to endure staying underwater for some time—perhaps a year, three years, or five years. So you want to gauge in your mind how long it will probably take and practice forbearance throughout this period.

What should we do during this period of waiting? We can live each day as it comes, one day at a time. And then, as we live each one to the fullest, we can cross it off on our calendar. It can get very difficult to think about what we have to accomplish during this time of one or more years of waiting, so instead we can think about each day as it comes and make the effort to use each day as meaningfully as we can. As you take each day one at a time with a calm and unshaken mind, doing your best to make each day as worthwhile as it can be, you'll steadily accumulate inner strength while you patiently wait for the tides of change to arrive.

◀ SEVEN ▶

Choose Your Best Life Partner

DISTINGUISH ROMANTIC LOVE from Marital Love in Your Relationships

The seventh secret is choosing the right partner in life, and this is especially important advice for young readers and those who want to get married in the future. Who you choose as your life partner plays a definitive role in the success of your overall strategy in life.

That being said, we aren't taught how to choose the right partner in life at school. Your parents might be able to offer some useful advice, but even they are not experts in marriage counseling, so they might offer you misleading advice—or they may be clueless about the subject themselves. Friends' advices are not the most reliable, either, because they might come from playful curiosity and amusement more than serious thought about what to look for in a long-term relationship like marriage.

The challenge of finding the right spouse doesn't allow much room for trial and error. We can't really go around test driving prospective partners; we can't exactly go through several different spouses first, just so we can make a better decision in the end.

In this sense, finding the right partner for your life can be a daunting challenge, but it is definitely a major milestone in the early stages of many people's lives—one that's as vital as choosing the right college to attend and finding your first job in real society.

Especially while we're young, our hearts are prone to romanticizing love and following wherever our fancy leads us. We can't hold back from idealizing popular men and women who always attract a lot of people's attention in the same way that movie stars do. So being in a romantic relationship during youth is exciting and can often consume our hearts with love's burning passion.

What we need to keep in mind while we're looking for a life partner, then, is that the romantic passion of our pre-marital relationships is different from the steady love we nurture throughout the long-term relationship of marriage. Romantic love and marital love are very different in nature, and we need to allow the romance to mature into marital love when we become committed to our life partner.

While we're single, we're more likely to think that we are supposed to look for a partner who shares our passion for romance. But in contrast to this expectation, a spouse is someone with whom we will be sharing many decades of mundanity and monotony. A partner with whom we can imagine spending our lives together in the long term and a partner with whom we can fall in passionate love may not always be the same person.

The reason these types of partners may not coincide is that we tend to be attracted to our opposites while we're young. Contrasting personalities and backgrounds are strongly alluring, and we become blind to their flaws and begin idealizing the positive qualities we find in them. So the stronger the contrast is between our personalities and upbringings, and the more strongly our friends, family, and mentors oppose our relationship, the more our feeling of attraction escalates. The more unfavorable our circumstances become, the more oil is added to our passion. This is often the nature of romantic relationships.

In comparison, marital love is quite the opposite. The time we spend with our spouse in marriage is full of monotony and is a long-term relationship that is nourished over ten, twenty, or thirty years of ordinary days and routine life. For example, it takes two decades of your life to raise your children before they become full-grown adults, and this is a process that is gradual. In this sense, the course of married life resembles the construction of a house. When the foundation is first laid, it's difficult to visualize how the finished outcome is going to look—but regardless of this, the construction of the foundation is the first and most crucial step to realizing a solid, lasting structure. It's after the foundation is set that the pillars go up, which are then followed by the walls, the roof, and finally the exterior. The process of marriage, just like the process of house building, needs to go through many gradual stages of construction that take time to be completed.

The process of marriage, just like the process of house building, needs to go through many gradual stages of construction that take time to be completed.

MEN AND WOMEN SEEK
Relationships Differently

Some men like to find fulfillment through short-term relationships, but a lot of women naturally gravitate toward the long-term happiness gained through lasting relationships. For this reason, women should know about the likelihood that a short-term relationship could result in unhappiness, because this type of love may not be what their hearts truly desire.

Men tend to go through many short-term relationships especially while they're young, much like moths that fly to and fro between this light and the next light. With little experience and wisdom to go by during their twenties, they will go through many trials and errors and will learn from making many mistakes. Since it's rare that they find a relationship expert to give them advice on how to succeed in relationships, it's very common for them to suffer a lot of failures. This is something that men should expect to happen to them to some extent. It is different for women, who are inherently capable of viewing relationships from a long-term perspective, and many women will find this natural tendency already lying within them, ready to help them create happy, successful marriages.

More women will be able to find true happiness by pursuing lives in tune with this natural desire. Many women in these modern times may be seeking physical relationships in hopes of finding meaningful love, but this pursuit often leads them to ignore the long-term happiness that their hearts are really seeking.

The way a woman finds happiness in a long-term relationship

closely resembles the way a spider finds nourishment. A spider's life is spent carefully weaving a delicate web between stalks. Patiently, it lies in wait in the center of the web for prey to become entrapped by its threads. It then sets aside the trapped prey as a source of vital nourishment. Over the course of many decades, a woman nurtures and develops her family and home, just as a spider weaves her web. And through supporting her husband, giving birth to children, and helping those children become successful adults, she is able to fulfill her innate desire to achieve long-term happiness.

One more point I would like to add is that intelligence and emotional maturity don't necessarily go hand in hand, meaning that it's possible for an intelligent person to fail in a relationship and for someone with a not-so-great record in school to build a very successful marriage with their life partner.

CAN YOU IMAGINE
Spending the Rest of Your Lives Together?

To return to the point I raised earlier about romantic love evolving into marital love when you become committed to a life partner, one important condition you should consider when making your decision of who to marry is whether you can envision you and your partner spending the rest of your lives together. Think about it, and consider whether you can see yourselves together ten, twenty, and thirty years down the line. And then ask yourself if you would like your partner to be the one to see you off at your funeral.

These are important questions that you want to have clear answers to, and if you find yourself saying no to them, then your relationship could be a passing romance rather than one with long-term potential. Someone who cannot imagine his elderly partner at his funeral is probably following the blind passion of a fluttering moth. You have to have the confidence to stay together through many years of each other's lives.

The most favorable scenario is to get married while your relationship is on the uphill climb toward the peak of romance, not when you've already reached it. Some couples get married at the peak of their relationship, or some might wait a few years after that in order to reach financial stability before tying the knot. But by the time that they are ready, the passion may have considerably subsided, and they may end up getting married out of obligation. It's hard to imagine a lot of joy arising from this type of scenario.

And an important piece of advice for when you start a marriage is to try not to expect perfection from your partner, just as you are not perfect. When you feel distressed and doubt whether you'll be able to make the marriage work, the crucial point to think about is whether you'll be able to share many years of your lives together under the same roof. The key to a successful life together is nurturing mutual understanding in your marital relationship.

CHOOSE A PARTNER
with Similar Intelligence, Physical Qualities, and Character

There are also several pragmatic aspects of marital relationships that I recommend thinking about when choosing a partner. The first piece of advice I would like to offer is to find someone who is compatible with your level of intelligence. It can be an essential factor for the happiness of your marriage to find someone with whom you can find some degree of connection on an intellectual level. When there is a wide gap between the two partners' interests and intellectual pursuits, the marriage may end up not working out in the long run. So I recommend finding someone who is generally considered by society as being compatible with you on an intellectual level.

For example, if someone with a brilliant mind were to marry someone gorgeous just for his or her good looks, they may eventually find it difficult to carry on a conversation at a deep level of connection. In this sense, choosing your partner because of his or her physical attributes can be an unhappy basis for an important life decision. This is a type of pitfall that young couples are particularly prone to, and it's a mistake we want to avoid making ourselves.

You'll have a much better chance at marital success with someone you can have a substantive conversation with, even if this person is not as physically attractive as you had hoped. A marriage is something that may last for your entire life, and in this sense, intellectual chemistry is a very important element of marital happiness. Physical beauty, no matter how exquisite, is subject to the hands of time, and the passage of ten years will show marks of decline. Physical beauty is something we can appreciate while we're young, but when it begins to diminish, we have to depend on intellectual chemistry to bring us happiness in the relationship. This is the reason why intellectual chemistry is one of the key elements of marital happiness.

The second piece of advice I have is another aspect of a happy marriage that you may rarely hear about, and it is about finding a partner with physical qualities that are a good match. It works much more favorably for a relationship if your partner's physical qualities, such as height, weight, and build are similar on balance to yours. A wide disparity in physical attributes can become a source of a lot of stress in day-to-day life at home. If a husband who's six feet tall is constantly teasing his vertically challenged wife, or if the wife is so much taller than her husband that he needs to constantly look up to talk to her, this may not seem a very significant issue while they're dating, but it may escalate into a problem when they begin living together on a daily basis. And the chances of a successful marriage between a very lanky guy and a professional female body-builder may be very low.

We would all like to marry someone who's intelligent, beautiful, or has other wonderful qualities, but what is more crucial is to look at how well your qualities complement each other. If you find a substantial difference in the degree or type of your attributes, it's probably best to reconsider the relationship.

Apart from intellectual and physical compatibility, the third key to a happy marriage is to find someone whose personality complements yours. This can be a very challenging aspect to get right, especially because many romances are sparked by contrasting personalities. The tension arising between opposing qualities can heat up the attraction the two partners feel for one another.

There are, of course, many relationships that work out between spouses with similar characters, but there are also romances that develop between extremely contrasting personalities. Relationships that fall into the latter category do have a chance of somehow being successful if, for example, the couple's strengths and shortcomings support each other rather than compete or fight with each other. But if they face a lot of butting heads, then the relationship may be difficult to work out in the end.

So, successful marriages are usually those between similar personalities. And if you're in a relationship based on opposites attracting, then what you should determine is whether your personalities help balance each other out and support each other's weaknesses. This will be a crucial point to consider in your situation. And if you find that your personalities are purely at odds with each other, then this relationship could be a bit difficult to hold together over the long term.

LOOK FOR SIMILAR Religious Beliefs and Values in Your Partner

A fifth key to a happy marriage is to find someone with similar religious beliefs. Religious differences can make it very difficult for a couple to spend many decades of life in each other's company, and this is why religion often plays a major role in marriage.

Probably the most difficult case we find is a marriage between someone with deeply religious values and someone of atheistic or materialistic leanings. In some cases in which both partners have faith in a religion, the marriage might succeed if their faiths teach similar values, but there are some religious beliefs that are inherently conflicting. A union between a devout Muslim and a pious Christian may pose quite a challenge, and couples in these situations should expect this.

Of course, love has the power to draw opposites together, and this may make them want to elope against the wishes of their loved ones. But these kinds of marriages often result in the couple's unhappiness down the road.

Having the same or at least similar beliefs and values is a very important factor in creating long-term happiness in marriage. So, determining your compatibility in terms of each other's spiritual maturity is an important way of foreseeing how well your partnership will work out.

Each person can generally be divided into one of nine stages of spirituality, beginning with the upper, middle, and lower stages, which then are divided further into upper, middle, and lower levels. If you and your partner are relatively close to each other in spiritual maturity—for example, if one is in the upper-upper stage and the other is in the upper-middle stage—then your values may be similar enough that you agree on many issues about life.

But it's likely that you'll face difficulties if you are too far apart from each other. For example, a marriage between someone from the upper stages of spiritual maturity and someone from the lower stages may fall apart at some point, even if both spouses are physically beautiful. Likewise, a devout believer and a hardened atheist may find it difficult to stay together, even if they were classmates in school. Even if they have something in common, such as sharing the same extracurricular activities in school, they may still find their hearts growing distant from each other as they lead separate spiritual lives. If you notice that this is happening in your relationship, you may need to accept the possibility that your marriage is not going to work out.

What this actually means is that you'll return to very different places in the other world. It's difficult to make marriages work out between couples who originally came from very different levels in the other world.

THE KEYS TO A SUCCESSFUL Marriage for Two-Income Couples

After we have determined whether we would be comfortable living with our partner, what are some key points we should consider in relation to our careers? There are two patterns that I have found to be traditionally successful in building a happy marriage.

The first pattern is that your partner understands the nature of your job and is capable of helping your work by taking on the role of your assistant. This pattern resembles the strong bond built from the friendship between two souls and can be quite successful.

The second pattern is that your partner is capable of independently managing all household responsibilities on his or her own. This partner might not have the specific skill sets to assist you at work but is capable of taking care of the family and home, leaving you worry-free about the domestic aspects of your life. In this relationship, your partner will be a vital support to you in your private life at home.

Either of these two patterns is highly likely to succeed and your choice is completely up to personal preference.

There are other possible patterns in marital relationships. For example, in some marriages, the husband may not have a fixed income, while the wife holds a successful job or career. Couples like this may have purposely planned this type of marriage while they were planning their lives together in the other world. Another scenario we commonly find is one in which both spouses hold jobs of their own. In both of these scenarios in which the female is working in society, the conditions for a lasting and successful marriage tend to become much more limited.

In these cases, the first condition for a happy marriage is ample financial wealth. You will need to be able to afford hired help and other items of convenience around the house. An ideal example would be if one of the spouses is a high earner with extra cash to afford these things. Another way this marriage can work is if other family members, such as the spouses' parents, are understanding of the situation and are willing to help take care of the children. Either of these two conditions will help make a double-income marriage successful.

But if the couple needs to rely heavily on both incomes to afford bare necessities, such as rent or a mortgage, this may be very challenging to work out. Couples in this case may need both of them to continue working until one of them becomes capable of earning enough to pay the rent on their own so that the other can stop working and stay at home.

A double-income marriage can also be more susceptible to extramarital affairs, so this is one advice I'd add as something we want to keep in mind when choosing a marriage partner.

◀ EIGHT ▶

Summary: The Seven Secrets to Attaining a Lifetime of Success

To sum up, I have discussed the seven secrets of becoming a successful leader, from setting the right aspirations to choosing the best life partner, and they are all based on the basic principles of happiness and success that I talk about in my books and lectures at my religious organization, Happy Science.

The shape of success that I constantly want to promote is the kind in which making the most of your gifts and capabilities leads to making the most of many other people's talents and abilities and succeeds in contributing positive work to the world. I believe that doing so invites the support of your guardian and guiding spirits of heaven as well as the help of friends and supporters in this physical world.

So when you run into problems and conflicts, you should examine whether your success is arising from self-centered desires. And if this is not the case but you continue to face difficulties at work, then you must wait patiently for time to be on your side.

Last but not least, your efforts alone are not enough if your goal is to truly succeed. You also need to employ the help of other people to assist your work. This help may come from the people who work for you or from the members of your family who wish to support you.

CHAPTER TWO

Developing Winning Strategies

Seven Keys *to* Continuous Success *in* Management *and* in Life

◀ ONE ▶

Look at Life as a Series of Battles

YOUR ACTIONS AND DECISIONS

Determine Your Defeats and Triumphs in Life

In this chapter, I would like to discuss how we all can learn to triumph over life's series of battles, which are a major aspect of our lives. Many readers may wonder whether the theme of life's battles is appropriate in the context of religion and spirituality. But as long as happiness and unhappiness, rises and falls, and constant changes occur in the lives of individuals, organizations, societies, and whole nations, we can say that each case represents a defeat or a triumph.

Ultimately, each defeat or triumph is the outcome of a series of actions and decisions we made during vital junctures in life when they counted the most. And this is true for individuals as well as for large groups and organizations of people. So in this chapter, I would like to look at life and this world from the perspective of defeats and triumphs and discuss the philosophy of leadership I have developed to help people triumph against their many battles in life.

We can begin by saying that life is a continuous series of defeats and triumphs. When we look at our lives, we'll only find a handful of major turning points, but on closer inspection, we'll discover that each day actually represents a battle of its own. Each day that we live, we need to check whether we have conquered or lost against what it represented in our lives. Put another way, each day is an opportunity to examine whether we have used it meaningfully—to consider whether the decisions we made, the philosophy we created, the conceptions we devised, the actions we took, and the relationships we built led to a triumph or a defeat.

> Ultimately, each defeat or triumph is the outcome of a series of actions and decisions we made during vital junctures in life when they counted the most.

Life is ultimately the collection of many of these days moving forward along a continuum, and the societies of this world are constructed by the countless individual continuums of people crossing paths with one another. This is a vital perspective for us to have on life and society as we live in this world.

We need to add an important point to this perspective, and it is that although a battle is an event that determines the victor and the loser, the defeated are also sharing in the rewards of the triumph. For example, the victory of an employee against a rival at work is determined by who wins the raise or achieves the promotion in the company. But from the company's perspective, the more workers there are who are winning individual rivalries, the more the corporation as a whole is triumphing against competitors in the market.

We can choose to believe that we need to be equal and work in conformity—that we shouldn't allow a select few to get ahead of others or that we shouldn't contribute ingenious ideas and productive work that shine above others. We can implement such a corporate system and culture that hinder select workers from emerging as victors. But we need to consider whether this is a system that could end up leading to the defeat of the whole. If we permit this kind of system, the entire organizational body may be forced to face defeat against its own competitors. The intention may be to prevent individuals from suffering from unhappiness, but this can eventually result in the whole corporation's failure and eventual bankruptcy, which will represent the defeat of everyone belonging to the company.

Therefore, it's for the benefit of the whole that there are exceptional employees who can come up with ingenious ideas, make exceptional discoveries, offer outstanding proposals, and produce new projects and work, even if they are shunned by colleagues and senior workers for doing so. It may appear as though many others are being defeated as a result, but the work of these exceptional people, from a broader perspective, is contributing to the company's overall success. The triumph of these people may seem as though it's causing many others to suffer defeat, but the truth is that everyone is ultimately sharing in the victory when the company becomes successful.

THE GROWTH AND DECLINE OF CORPORATIONS

Bring Benefits to Society

We can further consider triumphs and defeats in the context of intercorporate competition and examine whether one company's triumph truly results in another company's defeat. One company's success may signify another company's downturn in sales and eventual collapse, which does frequently occur in this world. These individual events can sound very grievous by themselves. But when we take a broader perspective of these events, we can see that corporate growth and decline is a process of the survival of the fittest that fosters the growth and survival of beneficial businesses.

What this does, in the end, is ensure the survival of good products and services that benefit consumers and end users. So consumers ultimately benefit from a system that encourages beneficial companies to survive and poor or harmful companies to decline.

There was a time when Japan's government-run postal service required us to follow the preestablished size requirements, and this posed a huge inconvenience to customers, who were the ones paying for these services. It was difficult to figure out the right measurements, which forced most people to resort to bringing their packages to the post office and being turned away if they got the measurements wrong.

Eventually, private companies began to offer package delivery services that offered pick-up services at customers' homes and offices without the hassle of size restrictions. What's more, these privately run delivery services offered one-day shipping. As a result, the government-run post offices stopped being as restrictive as they had been and began treating their customers more kindly and politely.

Eventually, however, their business declined and faced possible bankruptcy. This may seem like a very harsh reality, but it is a good example of how customers benefit greatly when companies with better services are able to survive.

COMPETITION FOSTERS
Growth and Development

At first glance, we find that competition produces victors and losers. On the other hand, competition is also a system that promotes the improvement of products and services and, in turn, the progress of society as a whole, and this is the reason we shouldn't disavow the practice of competition. When we make the mistake of doing so, we are fostering a world devoid of human work and effort.

A world where no one triumphs and no one suffers defeat, though it may have an attractive ring, will ultimately result in eternal stagnation for all. In such a society, all will necessarily suffer from the effects of defeat—everyone will stop aiming for further improvement and growth, and all will have to remain in a nonproductive state of torpor.

In this way, the sense of impending crisis—that we are never immune to the risk of collapse and ruin—works positively to promote corporate growth and success. In the same way, an individual's wariness of the risk of layoff helps encourage individual work and effort.

Developing a culture of many people who live and work in competition with one another may give rise to many individual tragedies in life or at work, but we can say from the macro perspective that, in such a system, all are aiming toward and progressing on the path of evolutionary advancement.

We, in this modern age, no longer need to fight battles by hand-to-hand combat as often as we used to, and in its place, we now engage in other forms of war, such as economic and academic competitions. We don't need to literally risk our very lives anymore. In addition, no one really loses in an economic competition. Those who rise above others as victors in this form of competition come up with ideas and conditions that are advantageous for many others, including those they have defeated.

When we think of how the shape of war has transfigured over the ages this way, we see that the competition of today is a system that really serves to benefit us all.

In the context of political elections, only one candidate out of all those running will emerge as the victor, and the rest will suffer defeat. Of course, if only one candidate was available to run for office, this candidate would be guaranteed victory in every election. So the current electoral system produces many defeated politicians, and this may sound like a sad condition. And it's true that some politicians end up depleting their assets to fund their campaigns. Some lose all their personal fortune, while others end up knee-deep in million dollars of debt.

On one hand, this is indeed a terrible tragedy for the defeated politician. But on the other hand, allowing more candidates to take part in the competition is a sure way of improving the chances that a better candidate will be put in office. And this is how fiercer competition allows us to create more favorable circumstances for society as a whole. It's not beneficial for the same politicians to repeat their terms in office only because other candidates aren't available.

COMPETITION GIVES Rise to Innovation

As you have seen from this discussion so far, the system of victory and defeat that surrounds us in this world today is a principle of happiness in disguise that fosters the principle of innovation. And as I have repeated many times, it spawns a beneficial state of society.

Long ago in times of war, battles were won by killing your enemy's general. The general's leadership was always needed at the time of battle, which meant that the side to be the first to kill the opponent's general would gain the victory and the side that lost its general would take defeat. And in many such battles, many high-ranking officers were killed in addition to the generals. As a result, those ranking below them advanced into positions of higher leadership each time someone was killed in battle. This system of advancement was the system of innovation in those days.

This was their way of using the principle of innovation in war back then, and today, this principle is based on a system of merit within corporate contexts—advancement is determined not by the death of superiors, but by the employee's capability and performance.

There are organizations with very heavy competition where superiors and subordinates are constantly exchanging positions. In some of the most successful companies, the section manager you used to work under as a freshman employee could become one of your subordinates. This is a common phenomenon we find in the majority of companies that achieve rapid growth. In an extreme scenario, you might even find a former vice president becoming your subordinate.

So the system of replacing higher-ranking leaders with those below them has evolved and is no longer based on death, but rather on a system of shifting and flip-flopping positions of leadership. This gives rise to clear wins and losses on the individual level, but it's the organizations that succeed in creating these systems in which the strong and capable win that survive and thrive in competitive industries.

LIFE'S PROBLEMS
Can Be Vital for Triumph in Life

I believe that it's essential that we hold in our minds this perspective of life as a series of battles. As individuals and as members of organizations, it's important to keep looking for ways to triumph against each of life's circumstances. Even if it's not possible to win every battle that comes our way, it's essential that we never stop searching for ways to evade complete defeat.

So when you are faced with a problem, you of course can search for the solution or try to determine what you need to do to find happiness or avoid unhappiness. You can look at your problem from these perspectives, but you can also think of it as a critical battle in your life that you must aim to win.

When you look at your problems and ask yourself, "What do I have to do to win this critical battle in my life? What judgment, action, and decision will lead me to victory and allow me to triumph in my life?" you will often find that you're able to think your problems through with a rational state of mind.

When we consider our problems in this way, we'll know that some level of damage is an inescapable consequence; we can't depend on finding the perfect solution that will let us avoid all loss. We'll need to consider this fact rationally and find a way to produce the best outcomes while incurring the least amount of damage. By approaching our problems from this perspective, we'll be able to narrow down our options and arrive at the best possible solutions. This is an important method for solving life's problems.

There may be times when we are unable to avoid the sympathy we feel for others. We may feel bounded by our relationships and unable to find a way out of the problem because of this. When we are faced with such situations, we can think of them as opportunities testing our ability to win and look at them calmly.

There is a move in chess in which a pawn is sacrificed to create an opportunity to take down the opponent's pawn. This is a difficult measure to take when the pawn feels valuable to us, but sacrificing it in spite of this can be the step forward that we need to take to address the challenges that lie in front of us. There are many approaches that can be taken to solve our problems, but viewing our problems as battles we need to face in life may be the rational perspective that we need.

So if you are someone with an especially sentimental leaning and you often find yourself in a knot trying to resolve your problems, sometimes you may need to step away from your ordinary approach.

You may benefit from thinking about what constitutes a battle in your life and what are the steps that you need to take to triumph against them. When you take this approach, you may be able to find a way out of the situation with surprising ease. This is an effective method of solving your problems that I encourage you to use when you need to face the battles of life.

◀ TWO ▶

Gain Foresight to Win in the End

SIDING WITH THE WINNER
Will Bring You Victory

When our aim is to achieve great results in life, we are bound to face situations in which we can't avoid making some sacrifices. We are likely to go through certain turning points when we need to abandon what worked for us before. In Buddhist terms, this is the practice of letting go of our attachments, and in military terms, this is the tactical strategy of giving up something we value for the sake of victory.

Sending a decoy to draw away the enemy's attention at the cost of losing the decoy is one such method, and so are feigning defeat and using sham tactics to trick the enemy. There are many specific measures like these that can be used, and what they all require is for us to accept some degree of loss and damage in compensation for gaining victory.

In the course of our lives, we all face major life decisions—such as selecting the college we want to attend, choosing our life partner, and choosing a job or vocation—that require us to choose from a large array of options. And when we do, we also need to consider how best to balance the public and private aspects of our lives—how to harmonize work life with family life at home.

There are many times in life when we need to make important decisions. For example, there may come a time when you'll need to choose between two supportive superiors who put in good words for you at work. Who you side with will be a critical decision that could hurt your career if you lack discernment and foresight. Or, if you work under someone who supported you in your career but a new superior with higher ability was just appointed, your ability to discern which of the two is likely to be successful will determine the success or failure of your own career.

In such situations, the key to your success is to choose the superior who will succeed. Choosing the one who will fail will lead to your own downfall. Your emotions may make it difficult to make the right decision, but when you need to make a major business decision like choosing a boss or a business partner, you need to be rational and be able to discern which of them will ultimately help your business succeed.

If we look at the outcome of World War II, we can say that Japan's defeat was actually a result of allying with the weaker side. The Japanese overestimated Germany's strength: they believed that Germany had the greatest industrial strength, when in fact it was not that strong. The Germans had just come out of defeat in World War I, and even though Germany had been showing an impressive surge of growth since then, it was still a country that had just recovered from defeat in war.

If someone in Japan with an eye for political and diplomatic foresight could have clearly foreseen Japan's slim chance against the Americans, even if the Japanese allied with the Germans, the Japanese would have avoided defeat. The only reason the Japanese joined the war was that they believed that siding with the Germans would bring them victory. Had they foreseen clear defeat, they would not have entered the war in the first place.

This piece of history shows how important it is that we do not ally with those who fail but side with those who are likely to gain victory.

LEADERS' DECISIONS
Must Lead to the Happiness of Many

We can also examine what happened in the Gulf War between the Americans and the Iraqis. If we look closely at both sides of the argument, we can find that there were justifications on both sides of this war. There are points of argument suggesting that justice was neither completely on the side of the Americans nor completely on the side of the Iraqis.

The Americans said that the liberation of Kuwait against Iraqi invasion was their just cause in the Gulf War. But a close examination of Kuwait will show that it wasn't a country that, by American standards, would have been considered admirable. The Kuwaitis were ruled by a small number of royals who held full control over Kuwait's oil resources and gripped over 90 percent of the country's wealth. From a democratic standpoint, Kuwait should have been considered an enemy requiring either reform or annexation by another country. Instead, the Americans chose to prioritize intervening against the Iraqis to help enforce international law in order to prevent further invasions of a country by another.

From the Iraqis' point of view, Kuwait was not only an awful, undemocratic country but was also taking large supplies of their oil from underground. The Iraqis had a point in feeling this way. Since their oil fields are connected underground, the Kuwaitis were technically depleting Iraqi supplies of oil by pumping larger quantities of oil than the Iraqis were.

Iraq was struggling to regulate and raise oil prices because of Kuwait's overproduction of oil. Setting a limit on production would have allowed the Iraqis to hike oil prices to a reasonable point and this would have allowed their national income to increase. But because of Kuwait's unhindered digging, Iraq was forced to watch its national wealth decline and its oil supplies diminish. Perhaps the Iraqis could have found some way to negotiate with the Kuwaitis through diplomacy. But, regardless of this, it was true that Iraq had a justifiable reason to invade Kuwait.

There are many aspects to how good and evil are determined. And depending on where we focus our attention among the diverse values and ways of thinking we find among individuals, companies, and nations, we can find understandable grounds for each side of any argument. There is rarely a dispute that doesn't have good reasoning on both sides. There is never a completely just side and a completely evil side.

If a neutral country had to decide who to ally with in this situation, it shouldn't take the side that would be defeated. If a country were to side with the weaker of the two, it would become the next target of hostile treatment. So if the people of that country felt that the Americans were likely to be the victors in that war, they would need to support the Americans. This is the kind of decision that's required from a nation's leadership.

So in the final analysis, those in leadership positions need to make decisions that won't lead to a large number of people's unhappiness. They need to think not with their emotions nor their personal, temporary values, but with foresight that can determine which decision will protect their people's lives and their nation's survival.

It's the same with those who work in a leadership capacity in a company or organization. In such a position, you may face the pressure to preserve your company's traditions and cultures and the founder's original spirit. The tradition of a company or the founder's philosophy is much like a country's constitution or the core teachings of a religion: it should be cherished, but it may need to be overcome and abandoned in times of life-or-death crisis, for there have been many companies that failed and disappeared because they adhered too strongly to old values.

MAKING THE WRONG Decision Can Lead to Defeat

Those in leadership roles need to always foresee the future of their country, company, or organization. They may face short-term battles, but to truly win, they need to be prepared for what lies in the future. Failing to do so and making the wrong decisions could ultimately lead to their defeat.

Important aspects of preparing for the future include the steps you'll need to take to ensure success and your ability to read the state of world affairs. You need to carefully examine your and your opponent's strengths, weigh the likelihood of victory versus defeat, and ponder whether or not this is your opportunity to succeed.

You need to carefully examine your and your opponent's strengths, weigh the likelihood of victory versus defeat, and ponder whether or not this is your opportunity to succeed.

In a sense, preparing for the future may account for 80 to 90 percent of your chances of winning. It's possible to foresee the outcome to some extent by analyzing the state of affairs and taking concrete steps to plan for the things that may come ahead. But when we try to go to war without doing these things, we most often end up walking ourselves into defeat.

To return to the example of Japan's defeat in World War II, the main reason for Japan's failure to win the war was its inability to continue fighting. The Japanese hadn't prepared enough to fight a lengthy war. They were prepared only to launch forays and surprise attacks for a short-term war, and they were planning to win it by forcing a peace truce.

They clearly had no chance against a country that was well-equipped to carry out a lengthy war, boasting great industrial strength and rich natural resources, especially if the war dragged on. The Japanese lacked preparation at the diplomatic level as well as in gaining a thorough understanding of the situation they were heading into, and this is how they failed to practice good foresight.

SURPRISE TACTICS
Work Only as Surprises

Generally speaking, when you fight a long-term war with many battles, the stronger opponent is usually the one that wins. There are times when the weaker side might win some victories, but this is similar to a junior wrestler who sometimes defeats a grand champion by coincidence. Such a victory doesn't guarantee that the junior wrestler is capable of winning consistently.

To have won World War II, the Japanese should have thought about how to gain continuous victories, not about winning isolated, short-term successes. And they made this mistake because of the confidence they gained from winning many victories in the Russo-Japanese War, which led to overestimating their own strength.

During this war, Heihachiro Togo used a naval maneuver called "crossing the T" to win a huge, decisive victory in a high-stakes battle against the Russian Baltic Fleet; this victory brought Russia to a crushing defeat. This success was so admired by the Japanese that they heavily relied on maneuvers designed to win quick victories, and this led to underestimating the value of preparation.

The "crossing the T" tactic that Heihachiro Togo used against the Russians was a maneuver that exposed the side of his fleet in front of the Russian warships. The warships were equipped with artillery not just on the front, but also on both sides and the posterior of the vessel. Since the firepower in head-to-head naval combat becomes limited to the artillery located on the front, Heihachiro Togo exposed the side of his ship, where there was more firepower, and successfully overwhelmed the Russians with a focused, superior attack.

This was a high-stakes maneuver, however, because exposing your sides to the enemy meant that your entire fleet could be critically damaged and destroyed if you suffered enough of their hits. So the fleet needed a strong formation and perfect timing to make this maneuver successful. What contributed to this success was the fact that the Baltic Fleet couldn't read Togo's intention, since this had never before been attempted in the history of naval warfare.

It takes a genius to come up with a surprise tactic like this. But the danger is that it does not work repeatedly. It's the same in the competitive world of business. If you are the challenger in a market with a large existing competitor, you will need the help of a surprise tactic that your competitor has never used before. It's almost impossible to make any headway using methods that have been used before, so you are going to need to do something very different in order to have a chance at victory. Meanwhile, your competitor will also become familiar with your methods. So it's difficult to tell who will be the victor once your competitor begins imitating your success strategy.

Surprise tactics sometimes enable the weak to defeat the strong, but we also need to be aware that these methods will be studied and copied eventually. It is better to save these methods only for when they are needed the most. We should take caution not to reveal them too often to our competitors to prevent them from mastering them. Because when the stronger and larger competitor masters these methods, that is the one who will take the victory.

◂ THREE ▸

Analyze Your Competitors' Strengths and Weaknesses

THE PRINCIPLE OF COMPETITION:

The Example of a Japanese Private High School

I have picked an example from war to illustrate for you that the approach of using surprise tactics carries the risk of leading to defeat if it's relied on too frequently. When we consider what this says about the principle of competition, we can say that those who take measures to succeed will succeed, while those who neglect to prepare will invariably suffer defeat.

It's possible to succeed by implementing a new successful tactic. But, even then, you will soon be followed by competitors who have the resources to replicate the same idea. At that point, it's a matter of time before they catch up and no longer let you remain at the top. At the same time, it will be impossible to avoid defeat as long as you cannot find the resources to adopt the same tactic. This is the difficulty of surprise tactics that is as important for young students to bear in mind while they plan for college admissions as it is for military leaders to consider as they strategize for battle.

A while back in Japan, a top secondary preparatory school called Nada High School succeeded in sending well over one hundred students to the University of Tokyo, one of the most prestigious universities in Japan, and became the school that sent the largest number of students there that year. The strategy that Nada High School took was to take advantage of its combined middle and high school system.

Nada High School instituted a new style of curriculum that allowed its students to complete their twelfth-grade education by the time they finished the eleventh grade. In other words, the students' curriculum was designed so they could complete their ninth-grade education while in eighth grade, their tenth-grade education while in ninth grade, and so on, so that when they reached their final year of high school, they could dedicate the entire year to preparing for the university entrance exams. Their twelfth-grade education became a college prep course.

Of course, other private schools quickly followed suit. They replicated the new system, gradually outrivaling Nada High School so that it no longer produced the highest rate of acceptance into the University of Tokyo.

While private high schools were using this innovation successfully and steadily increasing their rates of acceptance, public institutions weren't allowed to implement the new system. A rule made by the Ministry of Education prohibited them from doing so. As a result, many public prep schools suffered decline and fell behind in the college admissions competition, one after another. Even the most prestigious public prep schools saw a significant decline in their rates of acceptance and produced many students who failed to get admitted straight out of high school.

Many high school graduates in Japan choose to reapply to the college of their choice, and so they take college prep courses and retake the entrance exams the following year. But because of the greater number of private school students who were now very experienced at taking examinations, larger numbers of public school students were not accepted into top universities. In this way, many public high schools and their students were faced with hard defeat.

Seeing the success of the combined school system, the public school system started a movement to build new schools with a similar structure. But even if they do so, it's unlikely that they will succeed. The reason I say this is that the permission to accelerate their curriculum is essential to their success, but if the Ministry of Education gives them this permission, other public schools will also want the same permission. In my prediction, the Ministry of Education won't be able to offer permission only to the new combined school systems, out of a sense of fairness. If I am right in saying this, then it is unlikely that these combined schools will see successful results, and this plan may end up flopping.

To prevent this from happening, they need to assess and analyze the situation thoroughly as I have done just now so they can foresee the disastrous outcome before they implement it.

I would like to add one more point before we move on, and that is that my discussion is based on the premise that students who study in advance are at an advantage. But exceptional students who use a review-based method of study may benefit more from a slower-paced curriculum. The slower-moving curriculum will help them build more academic strength and will also have the advantage of complementing prep course curriculums. If you are such a student, you may want to consider whether going to a public school may be the better choice for you.

STRENGTHS AND WEAKNESSES
Are Two Sides of the Same Coin

A valuable aspect of preparing for the future is analyzing and understanding the strengths and weaknesses or successes and failures of your competitors. As we saw in the previous section, a preparatory high school should copy methods that have been successful in other schools, and the same suggestion applies to those who work for college prep courses. If a strategy or tactic is working for a competitor, then it is common sense to adopt it if you or your company has the resources to do so.

The weaknesses of your competitors can also teach you many important things. Strengths and weaknesses are often two sides of the same coin. For instance, if a leading prep course publishes a textbook believing that no one could possibly outdo such an outstanding book, it's not hard to foresee that other prep courses may try to analyze it and find ways to compete with it.

It's possible for someone to create a superior textbook to compete with even the most phenomenal textbooks by studying it carefully for weaknesses. You are bound to find some element or other that the textbook doesn't address thoroughly enough, and by focusing your material on this element, you'll find an opportunity to create a successful textbook. This is often called a niche-market strategy, and this is an approach that businesses commonly use to compete with other companies in their market or industry. And this is what business strategy is really about.

When a competitor brings out a good product or service, you need to find the right method or strategy to compete with it, because choosing the wrong strategy can lead you to defeat. There once was a prep course for high school graduates that chose a location near the University of Tokyo, close to the Komaba campus, hoping that the location would attract many customers to them. I think that this idea may have been successful to some degree. But if their customers found their teachers and materials lacking effectiveness and quality, then this prep course was probably outrivaled at some point.

Much like the feudal ages of long ago, schools and businesses of today are in a constant state of competition. Fortunately for us, when superior products and services score a win in these battles, their victories bring benefits to consumers and end users, and this is the reason we want to welcome competition with wide open arms. Of course, some businesses will suffer defeat and bankruptcy, which may tinge us with some sorrow, but this is a necessary sacrifice to foster the growth and development of businesses as a whole.

So paying close attention to the strengths and weaknesses or the successes and failures of your rivals is vital to preparing yourself or your company for the future. If you are enjoying a period of success right now and you see your profits continuously rising, your products consistently outperforming competitors, and the company consistently demonstrating results, you may feel confident of your ability to keep succeeding as long as you continue in your tradition or the methods that brought you this success. But you need to take caution, because this overconfidence could lead you to defeat when your rivals begin to copy your success.

ORIGINALITY IS THE KEY
to Continuous Success

So you will need to come up with a counter strategy for when a competitor begins to copy your success. Competitors might not imitate you while you're doing poorly, but they will begin appearing when your product or service becomes a bestseller or a long-time seller. This is something you should think about in advance, because it will be too late to do so once they start to appear.

The key is to have a secret ingredient that no one else will be able to identify. Like the mystery ingredient that gives a dish the special hidden flavor, your product or service should have that special element of success that no one is able to find out.

The reason why a product is so successful should be partly obvious to everyone, but part of it should remain impossible to identify. The quicker the secret can be found out, the quicker your success can also be quashed. The longer it takes for other people to discover the secret, the longer you'll keep your success going for yourself. But when the secret is found out, you'll no longer gain the same level of profit that your product initially brought you.

So when you develop a product or service that you know will become a hit, you'll need to predict when the imitations are likely to appear. You also want to keep in mind that you may need to accept defeat if your competitor has considerable capital. There are also cases in which competitors will wait and use you as their guinea pig first, and once you begin to succeed, they will siphon off most of your sales.

This is the very severe world of wiles and tricks that we need to face. Success is not just about a good product or service selling well and creating profit. There is also a world of intense competition that unfolds, and this is the reason that being aware of the strengths and weaknesses of your rivals is so vital to your success.

> The key is to have a secret ingredient that no one else will be able to identify.

This also applies to us as individuals. No matter what kind of vocation you are working in, whether you are a politician or entrepreneur, there is always a rival. If you are a student, you want to know who your rival is and examine what methods he or she uses to study and what kinds of weaknesses he or she may possess. You also want to find out what reference books and workbooks your rival is using and which prep courses he or she is taking.

Since each individual is unique, it's not guaranteed that you will succeed by imitating others, and this brings us to the fact that, when we try to copy others' success, we also need to determine which aspects of their success we should and should not imitate. In order to determine this, we need to also understand our own strengths and weaknesses.

◀ FOUR ▶

Concentrate Your Strength to Bring About the Best Results

There is a key to winning any sort of battle or competition we face in life, and that is to use the principle of concentration of strength. It is very important that we distinguish where we should use our time and energy to bring about the best results.

For example, if you are a student whose algebra grades are poor, you can improve them by increasing how much time you spend on practice problems and studying supplemental algebra books. In doing so, you're sure to see your algebra grades improve; the more time and energy you put into studying algebra, the more your grades will improve.

At the same time, however, if this strategy leads you to ignore your English, science, and social science courses, then your grades in those courses are likely to suffer. You are posed with the same dilemma as the battle leader; your flanks become weakened. You could hastily return to studying these subjects, but doing so will force your algebra grades to fall again. This is the dilemma we face when we concentrate our strength, and it's quite a challenging predicament.

Of course, what I have talked about so far is a method of reinforcing and strengthening your weaknesses, but there is another method of success, which is to reinforce your strengths—such as

English, for instance—to further enhance it as your strongest area of study.

You will see definite results from concentrating your strength. But your ultimate victory or defeat with this method depends on how you handle the problem when your flanks become vulnerable to attack. This is a dilemma that we find on the individual level, as well as in larger-scale cases such as a commander's strategy in battle. Ultimately, how you preempt the negative effects of this risk is the key to your triumph in life.

> Your ultimate victory or defeat depends on how you handle the problem when your flanks become vulnerable to attack.

If you are a student, you will need to decide how much time and energy you want to put toward your strong subjects, such as English, and how much energy you will dedicate to improving your weak subjects, such as algebra. Then, you also need to determine what you will do to prevent your other grades from falling. It's how you balance these three things that will be your key to success at school. It's about how you assess the outcome of the principle of concentration and identifying the point at which the results are plateauing. You need to recognize when spending any further time on that subject will reap minimal results and then shifting to a different subject to make better use of your time. This is the skill of dividing your time among your subjects for the most effective results, and your ability to do this will determine how your grades will improve over the course of one or two years of your education.

The same idea holds true with winning military battles. If you are a military leader leading a battle, concentrating your strength is about focusing your military forces where the greatest level of damage will be inflicted upon your enemy. Using surpassing force to assault your opponent's areas of weakness is a guaranteed method of victory.

For example, even an enemy of an enormous size, boasting a hundred thousand troops, is likely to have just three thousand men stationed in certain spots. These are the points you should determine to be your targets of attack. Sending ten thousand troops against these points will bring you certain victory. You can gain wins against them if you continue targeting their weaknesses in this way, one by one, in the divide-and-conquer method that Napoleon masterfully used.

Whether in business, in school, or at war, it is the forerunners, those ahead in their studies, and the countries with greater national strength that normally win. If you divide your forces evenly, the results are always going to be the same and as expected: the stronger army will win the battle.

So if you are the side with weaker military strength or the newcomer in a competition, the only way you have a chance against your adversary is basically through the use of localized assaults or surprise attacks. These are dark-horse tactics that can give the weaker side a good shot at defeating a much stronger foe.

This strategy of small, isolated attacks is effective and can help you gain a victory in battle. But while concentration of strength is an indispensable method of success, this tactic can simultaneously result in weakening your flanks. When you focus your energy on one point of attack, you automatically put yourself at risk in a number of ways. You risk an attack on your flanks, being assaulted from behind, and being surrounded. It can put you at a great disadvantage if your flanks fall under your enemy's attack.

So the question you need to ask yourself is, what will you do to preempt this predicament? The methods you devise to circumvent these attacks will be the decisive factors that lead you to rise to victory or crumble in defeat. I can't emphasize enough that the key that will make this method a success is in finding a way to defend these points or avoiding attacks on them, perhaps by keeping them undetectable to your adversary.

Once you have gained victories as the weaker opponent, you will need to think about how you will grow and transform into the stronger opponent. This is the next important but challenging stage in your path to success.

◀ FIVE ▶

Use Surprise Tactics When the Odds Are Against You

EXAMPLES OF Historical Victories

The use of a surprise tactic is one way that a dark horse has a chance of winning in a battle or competition against a stronger foe. We can find a fine example of this in Japanese history during the Genpei War, when the Minamoto Clan was on the verge of complete defeat at the hands of the Taira Clan, which was very close to conquering all of Japan. But a military genius, Yoshitsune Minamoto, emerged and turned the tides of war.

One of his notable battles was the battle of Ichinotani, a region located in the present-day district of Sumaku in the city of Kobe. Geographically speaking, the site of this battle was a narrow strip of land at the foot of very steep mountains, sandwiched between high slopes and the coastline. The Taira Clan had set up a large base in this location.

The Taira forces greatly outnumbered Yoshitsune's forces, spelling sure defeat if Yoshitsune launched his small army from the direction of present-day Osaka. So Yoshitsune devised a surprise tactic which he is now very famous for, the downhill attack on Hiyodori-goe. Traveling down from atop the steep mountains, he launched an attack upon the Taira clan's forces from their rear. The Taira forces hadn't foreseen the possibility of an assault from the direction of such steep cliffs and were caught completely off guard.

Yoshitsune's forces, which consisted of just a few dozen cavalrymen, rode down the steep slopes, infiltrated the base from the rear, and set fire to the central command. Seeing their central command on fire sent a shock throughout the enormous Taira army, and everyone fled in all directions.

And while the base was in panic and chaos, Yoshitsune simultaneously had a small allied army move in from the front of the base. This was how Yoshitsune succeeded at surrounding Taira's army on both sides, and so was able to prevail over the Taira clan.

> The use of a surprise tactic is one way that a dark horse has a chance of winning against a stronger foe.

Yoshitsune's strategy was a surprise attack and, in a sense, it was also a detour tactic. Instead of launching the attack from the front, Yoshitsune took a path around the mountains so he could attack the enemy from the rear, which allowed him to take the enemy completely by surprise.

This same strategy was used by the kingdom of Wei after the age of Zhuge Liang during the war of the Three Kingdoms in ancient China to defeat the strong kingdom of Shu. The Wei army marched through mountainous paths that no one would have thought of ever traveling through and charged down steep cliffs upon the enemy forces, just as Yoshitsune's calvarymen did later. The Shu forces were quite large, but the Wei army attacked the Shu forces' main stronghold, successfully forcing Liu Chan, the heir to the throne, to surrender.

So in summary, the Wei army took a detour, launched a surprise attack, and conquered the stronghold, defeating the powerful kingdom of Shu that Zhuge Liang and Liu Bei had formerly founded through brilliant military strategy and leadership.

We can find another example of a successful surprise attack in the famous Battle of Okehazama, led by Nobunaga Oda. Nobunaga was fighting the Imagawa army, an enemy ten times larger, boasting thirty thousand men. But because the enemy's front line was quite spread out, Nobunaga was able to find areas where men were taking breaks. In addition, it happened to be raining. So Nobunaga launched an attack straight upon the center of the skirmish line, successfully dividing the enemy's forces in two and swiftly killing their general.

Nobunaga's surprise tactic was very successful in this battle, but this was also the last time he used it. Surprise tactics are effective the first time they're used, because the enemy has no idea what is happening. But they normally lose their effect the second time around. Still, this is a very effective strategy for when you need to fight a larger enemy.

It's true that there was a second time that Minamoto Yoshitsune used a surprise attack: it was in the battle of Yashima. In Yashima, Takamatsu, the main forces of the Taira Clan had set up a base, and they controlled the seas in this area. This made the option of an attack by sea out of the question, and defeat seemed a certainty for Yoshitsune's forces.

So instead of going to Kagawa prefecture, Yoshitsune took a small troop of elite men by ship on a stormy night and sailed to the region of Katsuura located on the southern side of Tokushima prefecture, and there he anchored ship. Since it was too stormy at night for the Taira clan to patrol the seas, Yoshitsune's ship was able to make it there undetected.

So while the Taira Clan was expecting to be assaulted from the direction of the seas, Yoshitsune was scheming to invade by land from a different direction—from the south. Yoshitsune was able to gain the support of a number of the local clans in the area, and these clans did help him succeed, but the team of men that he initially took with him was very small.

In this battle of Yashima, Yoshitsune succeeded in forcing Munemori Taira to flee. And later, he brought the Genpei War to a final conclusion by winning the naval battle of Dannoura.

In this way, Yoshitsune twice used a maneuver to ambush the Taira clan from behind, both times catching them off guard, defeating their most vital military bases, and successfully bringing the whole clan to a collapse. The tactical genius that Yoshitsune demonstrated during many critical points of the war was truly extraordinary.

Today, someone with Yoshitsune's kind of tactical genius would become a successful entrepreneur with the ability to grow his business into a major corporation within his lifetime. He would probably start by vigorously growing his business in an uncharted area. Then, he would overtake the existing forerunners in the broader market, and finally, he would try to establish a stronghold in the market before other competitors could find a way to imitate his success. This is the course that his success would probably take if someone like him were alive today.

DETOUR TACTICS
Evade the Enemy's Anticipation

As you can see, there are two types of surprise tactics: one is to use a method of assault that evades the enemy's anticipation and detection, and the other is to approach the enemy by a detour path. We humans have the habit of trying to get to our destination by a straight course, so both we and our enemies are prone to using methods that can be easily found out by the other.

But using a detour tactic allows you to fool the enemy into believing that you are doing something else and succeed at reaching their base as if you have traveled through a multidimensional portal. This tactic camouflages your true intention, preventing your enemy from detecting what you are really trying to do. As long as you launch a frontal attack, you're making it easy for them to see through your strategy and tactics. But choosing a different route makes your objective and strategy unclear to them and encourages them to weaken their guard, during which time you have an opportunity to gather more forces to increase your strength.

People who think too simply use tactics that their enemy can easily discover, giving them the opportunity to prepare for an attack, so this is a tendency that we should caution against.

But one thing we should take note of when we employ a detour tactic is that we don't want it to fail and become a waste of a precious effort, and to avoid this, we want to hold a strong mental vision of our goal.

◂ SIX ▸

Create a System for Continuous Success

INDIVIDUAL VICTORIES
Are Often Short-Lived

To sum up, we have discussed five essential keys that lead to success in leadership: looking at life as a succession of battles, foreseeing the outcome, analyzing your enemy's strengths and weaknesses, employing the principle of concentration of strength, and using surprise methods of assault. After you have applied these keys, achieved some degree of success, and produced some level of results, the next key is to take to heart the fact that individual victories can be short-lived.

It's possible for individuals and organizations to achieve temporary success. But at some point, competitors are bound to emerge who will obviously begin to imitate your success, and this may alter the beneficial conditions that originally led to your success. You may have been fortunate to catch a tailwind that led your industry to hit the jackpot. But fair winds will abate and die out sooner or later, and competitors may catch up to you and surpass you at some point. There are also other possible factors that can lead your success to diminish or even lead you to fail down the road.

This is why it's important that you come up with methods that will continue to bring you success after you have secured some victories. Even if you happen upon a period of success, this normally will last only temporarily. So, preparing by building a system that continues to gain you victories in your personal life or your organization is essential. As with everything else in this world, individual victories are impermanent.

CREATING A CULTURE OF INNOVATION
Is Vital for Continuing to Win

Whether you are working for your own success or for that of your organization, resourcefulness is an essential element for continuous success. This means that you need to secure ample means and resources to allow you to keep fighting against competitors. For example, when you succeed at developing a hit product, you will need to think of ways to continue producing hit products year after year. You need to come up with new ideas that will bring you success in the next year, the following year, and as many as ten years down the road.

You may think that you are guaranteed to succeed as long as the same conditions are there or while you are keeping to tradition, but you may need to abandon tradition altogether at some point through a systematic disposal of the old. The conditions that brought you success before are not necessarily going to bring you success again.

A system of always succeeding is a guiding policy to follow, but it's not just that; it's also about creating a culture of innovation. It is about constantly being innovative, finding ingenious ideas, and revising and rebuilding your strategy. Both your individual life and your organization's strategy will differ tremendously depending on whether you have or have not aimed to develop a system that helps you keep winning.

We humans normally tend to just blame our good luck or bad luck for the victories and failures that arise in our lives. But those who don't do this, who live or work with a zeal to establish a system of continuous success are very strong. When conditions and circumstances change, the same tactics may not work as well as they used to, and we will need to adapt our methods to continue surviving and succeeding in a competitive world. This is what it means to create a system of continuous success. Thus, the key is to first focus on constantly thinking about how to keep succeeding. Because when you do, you will find new hints and wisdom arise from within you, showing you how to prepare for the future. By making the effort to think of an endless array of wisdom and gathering many people's wisdom to you, you will find your path to creating a system of continuous success.

When conditions and circumstances change, the same tactics may not work as well as they used to, and we will need to adapt our methods to continue surviving and succeeding in a competitive world.

◀ SEVEN ▶

Be Brave When Wisdom Runs Out

Wisdom and ingenuity can produce great success. But I would like to add one additional point to this, which happens to be a word of wisdom that the nineteenth century statesman Kaishu Katsu is also known for having said: There are times in life when we run out of wisdom and ingenuity, and when we do, our answer is to use courage. Kaishu Katsu faced many critical moments throughout his life. And while there were times when ingenuity came to his aid, there were others when wisdom failed to arise.

It takes a lot of wisdom and ingenuity, indeed, to be able to continue triumphing against each critical challenge we are faced with and to consistently remain the victors throughout our lives. It's for certain that there will be times in life when our fountains of wisdom dry up.

When you run out of resourceful tactics and schemes to help you succeed, the one last hope you have left to help you is courage. Courage is the final element to turn to when you need to win a battle or competition.

It's been said that Kaishu Katsu was the target of assassination attempts more than twenty times in his life, and in such turbulent times as his, there was no sure way of guaranteeing his survival. There were many assassin units constantly hunting enemies, many of whom were in hot pursuit of his life.

Such circumstances made it impossible to find any foolproof way of saving your life every time you faced a mortal threat. No matter how many security guards you surrounded yourself with, you were going to have to face death if someone broke through, much as Kozukenosuke Kira, the villain of the legend of the forty-seven *ronin*, was assassinated regardless of the heavy array of guards he hired to protect him.

So what did Kaishu do? He came out and faced his assassin unarmed. When he realized that yet another assassin was approaching him, Kaishu decided to show up without his sword, wearing his lounging vest. The assassin was dumbfounded when he saw him come out appearing as a friendly, good-natured old man. The assassin couldn't help but hesitate to make his move. It was one thing to assault someone armed, but it was another to kill someone with no means of self-defense. This was a kind of strategy that caught the assassin completely by surprise, and it worked.

Kaishu then invited the assassin into his home and sat him down for conversation. He showed him a globe of the world and gave him a thorough lecture on the current state of international affairs. The assassin's will to kill vanished, and he eventually went home. Legend has it that Ryoma Sakamoto, a legendary reformer and hero of Japan, was a similar assassin who initially went to kill Kaishu but went home as one of his converts.

Most people in Kaishu's situation would have brought their sword with them. But if the assassin had been armed with a pistol, or if there had been a whole army of men, the power of a sword would have had no chance against them. There are some situations in life when your only, final option is to wield your courage.

This works not only for individuals but also for larger battles in war. There is some debate about whether the last shogun of the Tokugawa Shogunate, Yoshinobu Tokugawa, was truly a great leader. Some say that he was the greatest leader since Ieyasu Tokugawa or that he was the Ieyasu reborn. But, in my opinion, he was prone to depending too much on knowledge and lacked courage when it was most needed.

Intellectually, he was strongly influenced by the Japanese Mito tradition and philosophy of showing allegiance to the emperor. So when his enemy gained imperial backing, he became incapable of winning the war. He was an exceptionally intelligent man who had deeply studied this philosophy, enrooting it into the depths of his mind. And when his enemy was incorporated into the imperial army, this event became the cause of his demise.

Perhaps his defeat worked out for the best of everyone. But I believe that there must have been other possible outcomes. Yoshinobu had the larger army, which gave him the advantage, so he probably could have found many ways to avoid defeat if he had practiced better leadership. In the final analysis, I believe that the real reason for his defeat was simply that he lacked courage.

As they say, too much scheming can become the schemer's downfall. Likewise, intellect can become the downfall of the deeply intellectual when they reach the limits of their intelligence. There are times when we need to turn to the wisdom of resourcefulness for a better solution, and if that doesn't help, then we need to look to physical and mental endurance. Resourceful ingenuity can produce a hundred-fold to a thousand-fold and perhaps even ten thousand-fold the power of physical endurance.

But there are times when we run out of resourceful ingenuity, too, and when we do, our final option is to have courage. There are points in life when we all have to make various high-stakes decisions, whether in our personal relationships with others or in our corporate investments—for example, in developing new strategies or planning an overseas expansion. And no matter how much we collect information, gather research, and ponder what we should do, we may not be able to arrive at a conclusion.

> When we run out of resourceful ingenuity, our final option is to have courage.

When we find ourselves in this predicament, we cannot remain undecided this way. Sometimes, life calls for us to make a decision even if it turns out to be the wrong one, especially if we are in the position of a top executive of a company or organization.

So I'll repeat again that we need to use our courage in the end. If our decisions turn out to be wrong, we can still find ways to amend the course as we go. But as long as we don't make a decision at all or we keep putting it off, we won't have a chance to find out what was right and wrong. The right answer becomes clear the moment that we realize we made the wrong decision. And this can help us think of a new strategy or tactic, whether that is to retreat our forces or to use some other method of fighting our enemy.

In conclusion, another key to winning the battles of life is to have the courage to decide on a course of action when our wisdom and ingenuity run out.

CHAPTER THREE

Cultivating a Management Mentality

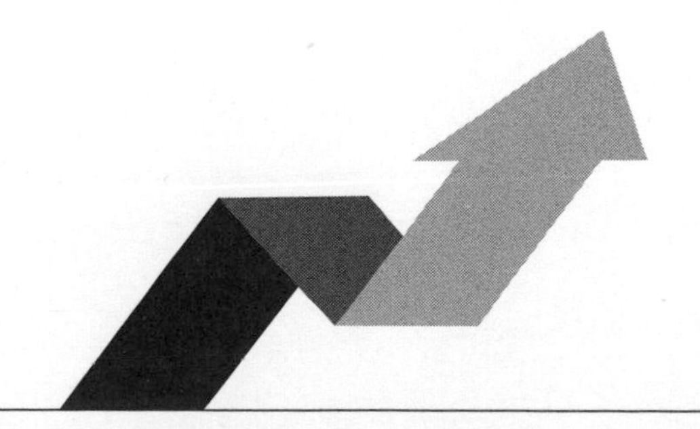

Five Keys *to the* Development and Prosperity *of* Your Business

◀ ONE ▶

Keep On Thinking

KNOW THE Purpose of Management

Developing an effective management mentality can be quite beneficial not only for those who are in managerial positions, but also for anyone who interacts with others in their day-to-day work. What I mean by "mentality" here is our way of thinking, including our mindset, mental attitude, and intellectual capability. So a management mentality is the way managers think.

Apart from those who work completely alone, all of us, at some time or other, have worked or will work above or under somebody else. As soon as we have people working under us, we are expected to fulfill the role of manager. Likewise, understanding how managers think can be of great benefit to us when we are working for others.

So in this chapter, I would like to discuss the mental attitude we need to cultivate to become top executives, which will essentially be about how we should go about cultivating and developing our mindset through the course of our lives.

In the modern world, where everything constantly grows and develops, being content with the current state of affairs and staying the way we are will inevitably lead our company to stagnate, lose out in competition, and eventually disappear. To survive in this evolving society and to support the livelihood of its many employees, an organization needs to constantly aim for development and prosperity.

Ultimately, the purpose of management is to further the development and prosperity of the organization. Those of us who are aiming to become effective executives or managers need to always cherish visions of how our business will develop and prosper and devote ourselves to actualizing them through our day-to-day work.

This idea of creating business development and prosperity is part of the principle of progress that I teach at Happy Science, which, along with the principles of love, wisdom, and self-reflection, constitutes the four principles of happiness.

PRODUCE
Something New and Better

The first essential key to cultivating a management mentality is to keep thinking. This may sound almost trite, but this simple practice will lead us to the path of development and prosperity.

The world in which we live and work is in constant change. Nothing remains unchanged. Everything keeps changing and evolving day by day. We are influenced by things that happen around us, including the way our competitors conduct business and the products they sell as well as what the market wants, seeks, or likes.

Our mental endeavor to keep thinking is essential for us to continue working in this ever-changing world. We would not be able to successfully manage a business if we became fixated on a decision we made three, five, or ten years ago and refused to change it no matter what happened.

Communist and socialist nations were often governed based on an archaic ideology in which all top decisions were final and never overruled. In fact, some organizations still operate under such rigid and inflexible management policies and principles. For example, local government offices in Japan still follow rules that were set before World War II.

Under a free economic system, however, businesses need to make innovations not on an annual basis, but on a daily basis to cope with changes in society. We want to avoid making drastic changes to the company's important decisions too often, as doing so could impede business operations. But making improvements in our daily routine work is essential to achieving the company's fundamental mission. Innovation is about coming up with something new and better, something that didn't exist yesterday. And to do this, we need to keep thinking. Continuing this mental endeavor will help us cultivate a management mentality.

Much of the time, our minds are occupied by random thoughts that arise in response to the people and things we encounter throughout the day. But this is not the type of thinking I am referring to. To think is to concentrate the mind on a specific idea for a certain period of time. It takes a lot of effort to make a habit of thinking things through.

Although this may sound simple, it's actually not so easy to practice, because the way we think has to be appropriate to our position in the company or organization we belong to. An office clerk would probably have a hard time thinking the way the CEO does.

Thoughts have power. What really matters is not only whether you keep thinking, but also whether you produce any results or outcomes by thinking things through. It is the process of thinking and creating something new from them that matters the most.

FIND THE NEEDS
Rather Than the Seeds

I myself constantly come up with lots of ideas and receive a variety of inspirations from heaven. But I can't always make use of these ideas straight away in the management of my organization, Happy Science, because the inspirations I receive and the ideas that I think of are often not the type of ideas I need for management.

We often receive inspirations about the fields or topics that we take a personal interest in, so they tend to be personal and specific to each individual. Those who are artistic or scholastic in nature are apt to receive ideas and inspirations, because these types of people are always looking for *the seeds*, or subjects they can write about. Scholars are seeking topics for their dissertations and research studies, and writers are constantly looking for a subject that they can develop into a story.

Artists, novelists, scholars, and those who work on their own can certainly benefit from receiving unique ideas, because they can produce good books and paintings based on the inspirations they receive. They can make use of the seeds they find and develop them into work to make a living and to support their families.

But finding the seeds is not good enough for top executives of an organization. They need to know *the needs*, not the seeds. They have to figure out and provide what people are seeking. So, unfortunately, unique ideas within the range of their personal interests and curiosity won't necessarily help them in their business operations.

Those who are in charge of managing a business of a certain size—with a workforce of anywhere from fifty to ten thousand employees—are responsible for running their business with its own set of goals and a unique corporate culture and for providing products and services to customers. In this respect, the ideas that top executives come up with have to satisfy the demands of the market and at the same time suit the purpose of the business. And those in management positions need the skills to use the people in the organization to implement their ideas.

The ideas that top executives come up with have to satisfy the demands of the market and at the same time suit the purpose of the business.

REFINE YOUR IDEA UNTIL You Can Develop It into a Profitable Product

One of the objectives of any business is to make a profit. A business's earnings provide for the livelihoods of its many employees and support the cost of its future development. So it is also the job of top executives to think of ideas for products or services that will generate profits, for any product, service, or project that does not bring profits to the company will sooner or later be eliminated.

If you started your own business, you must have a wealth of ideas. But to develop your ideas into something that can generate profit for your business, you first need to narrow your ideas down with a specific purpose in mind.

Artists, writers, and scholars often practice *divergent thinking* to come up with many different and unique ideas that no one else has ever thought of. They try to broaden the range of their ideas. A wealth of ideas can certainly be a valuable asset. But managers and executives need to practice *convergent thinking* to focus their thoughts and to concentrate their minds on a single objective. Those in management need to narrow down their ideas and consider whether their ideas will be practical, useful, and implementable.

As a business executive, you need to examine and reexamine your idea repeatedly until you develop it into something that you can use in your business. While going through this process, you can filter your ideas; you may sometimes find that some of your ideas come to fruition while others do not.

You can certainly focus on your ideas and think them over in your head, but to avoid finding yourself building castles in the air, you can perhaps write your ideas down or discuss them with others, preferably those with management skills.

Even when you feel that your idea is ready for practical application, you should still share it with your friends and colleagues and ask what they think about it. You may learn that a similar idea failed in the past or discover some drawbacks to your plan. This way, you can toss around your idea and refine it further.

Think of this process like a conversation between Sherlock Holmes and Dr. Watson. They solve difficult cases by throwing a conversational ball back and forth between them. We can use a similar approach in business management to hammer out a plan: one person proposes an idea, and another takes it and gives feedback.

If you are running a private business, you may be able to make all the decisions by yourself. But as your company grows, it becomes increasingly difficult to implement all your ideas. So try to find at least one, two, or, if possible, several people you can consult with. Share your ideas with them, ask for their feedback, and watch their reactions. As you keep the conversation ball going, you will be able to gradually refine your ideas. Repeat this process until you hammer out a concrete plan.

The situation may differ in the case of a major corporation with thousands of employees. The process for a large corporation is more formal and complex than the process for a small company. Often, dozens of board members vote and reach a decision in a process very similar to the process the government uses to make decisions.

If your business is a smaller organization, you may not be able to hire many managers. But you should still be able to find a few people with whom you can discuss your ideas, iron out any drawbacks, and work out the details.

So the first key to cultivating a management mentality is to keep on thinking. Once you think through your ideas, you need to discuss them, and exchange opinions with your colleagues or other managers to decide whether your ideas are practicable in your business. Go over your ideas repeatedly until you can refine and develop them into products or services that can generate profits for your business.

◀ TWO ▶

See the Big Picture as Well as the Details

BROADEN
Your Business Perspective

The second key to developing a management mentality is to see both the forest and the trees.

Grasping the big picture and understanding the overall situation are essential skills of an executive manager. For instance, let's say that you've developed a unique product using your special skills, and it becomes a big hit. As soon as the business grows, you will no longer be able to handle everything on your own, and you'll start hiring people. Once your business reaches a certain size, you will inevitably be confronted with management decisions that you may not be able to solve using your technical skills. This may make it difficult for you to keep running your business.

As an executive manager, you need to gain a broad perspective to see the big picture. It becomes essential that you observe the overall situation of what's happening around you—in your company, the market, the domestic economy, the world economy, and politics—and identify emerging global trends. This is an extension of thinking through our ideas. You need to keep gathering new information from the outside world to be able to know what it is that you have to do.

You need to consciously develop a broad perspective to become an effective executive manager who is responsible for leading a large number of people. It is the manager's job to come up with bold and innovative ideas for the entire business beyond the scope of the day-to-day work. Managers need to think about the company's overall operation, its relationship with the outside world, and its future, while those who work under them can concentrate on the work at hand.

It is the manager's job to come up with bold and innovative ideas for the entire business beyond the scope of the day-to-day work.

Whether or not you can see the big picture has to do with your aptitude to a certain degree, but it is an ability that needs to be developed, or it will go unused. The starting point is to be mindful of the importance of constantly cultivating a wide perspective.

KEEP AN EYE on Small Changes

Seeing the big picture is not the only ability we need to cultivate to become successful executive managers. We also need to pay attention to small details—the little things that tend to be overlooked. Doing this will open up new opportunities for the future.

Sitting in the executive's office and signing documents do not qualify someone to be a manager. Some corporations and government offices in my country, Japan, maintain a seniority system consisting of ten to fifteen hierarchical ranks into which employees get promoted not because of their capability, but simply because of the number of years they have worked for the organization. People who rise into a high position this way often mistakenly believe that they are grasping the big picture because they stand over many others and countersign the documents that have been passed up to them from their subordinates. They may be able to stay where they are as long as the company is prospering and developing, but I don't think signing papers is the skill we need to cultivate to survive in the hard economic times that we are currently facing. To become effective executives, we need to constantly pay attention to small changes in our organizations.

MAKE DISCOVERIES through On-Site Observation

How can we become mindful of small changes while grasping the big picture? The answer to this question may be summarized in a phrase "on-site observation": it is to visit the factory, store, or location where our main business takes place and see what's happening there. We may notice things that the people who work there do not. We may also discover that the business is being run in a way we never expected.

Let's say that you are an executive manager of a nationwide

supermarket chain and you are making management decisions based on reports you receive from locations around the country. One day, you decide to visit one of the stores, and you find that the actual situation is completely different from what you expected. The sales floor is arranged in a way that's inconvenient for the shoppers, the cashiers are behaving rudely, and the staff is ignoring the customers' complaints or too busy chatting among themselves to notice someone who needs assistance.

You also find out that store clerks are hiding the best-selling products at the back of the shelves while placing the slow movers in the best spots. And when you ask them why they do that, they tell you that they place products that don't sell at the front because they want to clear inventory. They hide the best sellers behind them so that slow sellers will catch people's attention. As a top manager, you see the problem right away and realize that this practice will damage your business. It's only natural that customers will complain if the store displays what they do not need and hides what they do need.

While overseeing the entire operation of the company from a broad perspective is essential, we also need to know what is happening in individual stores. In the case of the supermarket, you, as an executive manager, need to know everything from the sales situation, the size of each store, the products the stores are selling, and the services they provide to the attitude of the sales clerks and the reactions of the customers. You also need to make sure that every complaint is handled with care, because today's complaints are the seeds of tomorrow's success.

Likewise, if you are a hotel manager, you are not doing your job if all you do is sit in your office and sign the papers that come to you. You need to visit the hotel and see for yourself, for instance, whether the elevators are working properly, the stairs are clean, the design of the wallpaper is suitable for the hotel, the porters are courteous to the hotel guests, and the staff is providing good services.

Observations of small details like these can significantly impact the overall operation of the company. A small discovery that you make in one location can help improve services at all other locations.

To develop a management mentality, you need to be able to see both the forest and the trees. To broaden your perspective, you need to keep studying and learning new things by reading books and newspapers and watching the news. This will enable you to see the trends of the times and the future prospects of society as a whole and of the industry you are in. At the same time, you have to pay attention to the small details of your business. Both broad and detailed perspectives are essential for effective business managers.

You may be wondering how you can simultaneously see things from both of these opposing perspectives, but this is something you have to learn through practice. You will be able to get the knack of management by striving to see the big picture while paying attention to the details.

◀ THREE ▶

Discover and Solve Bottleneck Problems

FIND THE BOTTLENECKS that Hinder Business Growth

The third key to developing a management mentality is to find and unblock bottlenecks in your business. As the name suggests, "bottleneck" refers to the top, narrow part of a bottle. When you pour water from the bottle, the bottleneck limits the amount of water that comes out.

A similar phenomenon occurs when an organization grows to a certain size: it runs into bottlenecks in the process of development. Just because the organization is big doesn't mean that it can handle issues of all sizes. Anything that everything or everyone has to pass through—department, procedure, place, or person—can become a bottleneck for the organization. Wherever bottlenecks are, they hinder the further growth of the company.

As a manager, you have to know that your business will, at some point or another, come up against bottlenecks as it develops. Even when everything else is working well, a single bottleneck can hamper the growth of the whole business. Discovering and fixing the bottleneck becomes the key to the business's further development.

ADJUST YOUR THINKING
to the Size of Your Business

Let's say you are a manufacturer and you have recently expanded the size of your factory to increase your production output. This itself is great news for your business. But without a sales network that can sell the additional products, you will end up with surplus of stock in your warehouse. In this case, the bottleneck of your business is an insufficient sales and distribution network. If you don't have enough space in the warehouse to store the products manufactured in your expanded factory, a lack of storage space is the bottleneck.

Even with a large sales network across the nation, your products may not sell at all if you lack trained sales personnel. In many cases, good products do not sell as expected because of a lack of proper training for sales representatives.

Sometimes the bottleneck can be the inability to recognize the need for certain departments that become essential as the business grows. If you started your business with a launch of a new product that's selling well, you may only be concerned with how to improve the product or how to boost sales. But as the business grows, certain departments become essential for the company to function. To someone who initially handled everything from personnel affairs to bookkeeping, departments that specialize in general affairs, accounting, and human resources may seem unnecessary and may even seem like a waste of labor expenses.

But these departments are essential for the company's continued growth. If the business owner can't let go and tries to do everything by himself, he will not be able to do his own job, whether it is going out to sell more products or concentrating on the research and development of a new product.

The bottleneck that hinders a company's growth changes as the company develops. So, to find and solve the bottleneck, you need to change the way you think according to the size of your business. Even if you solve the problem that stands in your way now, you will soon face the next hurdle. Discovering and overcoming each bottleneck will open a new path to the further development of your company.

To find and solve the bottleneck, you need to change the way you think according to the size of your business.

LEARN FROM COMPANIES that Are One Step Ahead of Your Company

Another barrier to business development is the executive's inability to change his or her thinking to keep up with the growth of the company. In most cases, owners of small to medium-sized businesses do not realize that they have reached the limits of their capability and stick to old-fashioned ways of running their businesses.

So if you want to run a growing business, it's essential that you develop yourself by gaining and learning new knowledge. The best way to do this is to study companies or organizations that are one step ahead of yours. This is how you can see the future of your company.

Let's say that you are the owner of a business with one hundred employees. By studying a company with three hundred employees and learning from them how they operate their business, you will be able to see where your company will stand a few years from now. If you manage a company with a workforce of three hundred people, you can observe a company with five or six hundred or even a thousand employees and see how they carry out their business. Waiting until the last minute to start thinking about what to do may jeopardize the operation of your entire business. Select a company of a size that you want to aim to reach in ten years. By the time your company grows to that size, you will be all prepared; you will have already known what it is that you need to do.

As an executive manager, you need to constantly envision the future of your company and solve any bottleneck problem that stands in your way. To avoid becoming a bottleneck yourself, start preparing yourself ahead of time. Think about what it is that you need to do to bring the best out of your employees and to build an effective organization. This is a difficult but a necessary task that you need to take on to develop your business to the next stage.

◂ FOUR ▸

Always Be Mindful of Customers' Needs

PUT YOURSELF IN the Customer's Shoes

The fourth key for cultivating a management mentality is to always put yourself in the customer's shoes. I'm sure you've heard this saying before, but it's worth mentioning it here because we can easily forget how important this attitude is.

Because manufacturers and producers focus on creating the products, they tend to see things only from that perspective and overlook the needs of their customers. I understand that manufacturing is a difficult job. But when their products don't sell, they often blame others. They may say that it's the consumers' fault for not buying the products or the government's fault for not providing sufficient funds for the development of the product. They may also attribute their failure to other companies, other countries, and the fluctuating currency.

The manufacturer of the product may claim that it is the best in the country, but whether that is true is up to the market to decide. Customers make the correct assessment of the product. Just like the approval rating of the government averages out throughout the country, our sales capacity will be measured by how well our products sell on average, regardless of where they are sold.

The market is the total sum of the consumers, so it represents the opinions of all the customers. If your competitor's products sell more than your product, it means that consumers chose their products over yours. Ultimately, manufacturers have to ask the market why.

To find out why customers bought their products and not yours, you have to put yourself in their shoes and think. The answer to this question will become the seed of your company's next innovation, because it will tell you what improvements you should make.

DEVELOP NEW IDEAS from the Consumers' Viewpoint

To understand the customer's needs, we need to develop a customer-first attitude. This attitude should be adopted not only in for-profit businesses, but also in non-profit organizations. For example, the Japanese government went through an administrative reform in 2001. They restructured government ministries and agencies and significantly reduced the number of governmental offices and officials. But they did this based on their own needs, not considering whether it would benefit the people who use the public services. People should be able to decide which offices and services to keep and which ones to eliminate.

The government's customers are the people who use their services. So to find out which services are necessary or unnecessary, government officials need to directly ask the citizens of the country who rely on public services. They have to visit local offices and ask the people there what it is that they need to adopt a consumer's perspective.

Even top officials in the government can't figure out what the people need if all they do is sit in large, comfy armchairs reading newspapers. Politicians won't be able to tell what measures to take unless they see with their eyes what's going on and listen to the people who use government services in local offices. They can very well make wrong decisions if they only follow their own guidelines. To make the right decisions, government officials need to see things from their customers' perspective; they should put themselves in the shoes of the people who use their services.

Top executives of for-profit businesses can make the same type of mistake. Those at the headquarters often make important decisions based on the figures in the reports they receive, but if they completely rely on these numbers, they can make grave mistakes.

Prices and demands are up to the market to decide. We can in fact learn this from the mistakes of the past. One of the reasons socialism collapsed was that the Politburo did not understand that commodity prices fluctuate and that the government cannot control the market. The party leaders set prices and production quotas ignoring various interrelated factors that affect the market prices of the products.

Top executives need to be keenly aware of the thoughts and feelings of the end users of their products and services.

A socialist system isn't something that only existed in the Soviet Union or China. Organizations all tend to incorporate a similar system when they reach a certain size. As a matter of fact, major government offices in Japan practically operate under a socialist system. Large corporations are prone to becoming bureaucratic, too: they often follow rigid rules and do not decide anything until they go through an approval process that involves people of many different ranks.

We may make the same mistake that the communist and socialist nations made if we mistakenly believe that we know what's best for our customers because we know better. How can we avoid this pitfall? By putting ourselves in the customers' shoes. We should always try to see our products and services from our customers' perspectives and study their needs. Put simply, top executives need to be keenly aware of the thoughts and feelings of the end users of their products and services.

◀ FIVE ▶

Increase the Added Value of Your Products and Services

ADDED VALUE

Is the Sum of Customer Satisfaction

The fifth key to developing a management mentality is to increase the added value of your products and services. Let me first explain what added value is.

All business or commercial transactions involve products or merchandise, which could be anything from goods to ideas or programs. You also need a distribution system for your products to reach your customers. Finally, you need to provide customer services to people who are interested in your product. The added value is the total sum of customer satisfaction with these three elements—products, distribution, and services.

Let's say that an electronics appliance company released a new model of refrigerator that the engineers and designers feel confident is the best refrigerator in the country. Even if it indeed is a top-quality product, for this refrigerator to increase its added value, the customer also has to be happy with how it is distributed and with the service she receives at the store she purchases it from.

The refrigerators need to go through different channels to be delivered to the retail stores. If something happens during this distribution process, you will receive complaints from your customers, such as, "It takes too much time for the product to arrive when it's on back order" or "I can't get the replacement parts when the product breaks down." In this way, problems in the distribution process could hamper your efforts to sell your product.

> The added value can be gauged by the sum of your customers' satisfaction levels, which depend on the quality of the product, the distribution operation, and customer service.

Even if the refrigerators arrive safely at the retail stores, they still may not sell if the sales clerks don't provide good customer service. Customers won't buy the refrigerators if all the clerks do is simply place them right next to all the other models of refrigerators or, worse, if they have a bad attitude, as if to say, "Buy it if you want it." Poor customer service could ruin all the effort that the design and production team put forth to produce the new refrigerators.

To increase sales, store clerks need to provide valuable information about the product. They can explain why the latest model is better than older models and what improvements have been made. They can show the customer various features of the new model—such as energy efficiency, ability to maintain a steady temperature, and a high-performance freezing function—and explain how it is a good deal considering all the factors, including size and price.

A good product could lose its value if we don't provide comprehensive services all the way to the end user. A rigid bureaucratic structure in which executives ignore customer complaints and instead simply push people to buy the company's products will only decrease the products' added value.

To sum up, increasing the added value of your products and services is essential to produce a longtime seller that will earn profits for your company. And the added value can be gauged by the sum of your customers' satisfaction levels, which depend on the quality of the product, the distribution operation, and customer service.

CREATE A CORPORATE Culture that All Employees Share

If you own a small store, you can probably handle all aspects of your business by yourself, but in a large organization, it is impossible for one person to handle everything from manufacturing the product to delivering it to the end user. In a big corporation with tens, hundreds, or even thousands of employees, various groups of people are assigned to different tasks: one group manufactures the product while another plans the sales strategy; one department handles distribution, and another department is responsible for selling products in stores.

For a corporation with a complex structure to achieve its goal, employees need to share a corporate culture that helps them join hands and work together toward the same goal. All the parties involved in the process of creating, distributing, and selling the product need to be keenly aware of what they are supposed to do.

For the organization to keep growing, it is essential to raise the awareness and morale of the entire staff so that they are actively thinking and motivated to work. Everyone, from the CEO to store clerks, should be focused and proactive about solving any existing problems and eager to improve the company's overall performance. Not just the top executives, but the entire staff should keep thinking, with both broad and detailed perspectives; try to eliminate any bottleneck problems that stand in the way, and take good care of the customers.

We can enhance the overall capability of the organization when all employees share a corporate culture. The management's job is to build this culture and let it permeate the entire organization. We can do this by repeatedly sharing the philosophy behind the corporate culture over and over again so that all members of the organization share the same sense of mission.

◀ SIX ▶

Summary: The Five Keys to Developing a Management Mentality

To conclude this chapter, let me review the five essential keys to developing a management mentality.

THE FIRST KEY is to keep on thinking. Focus your mind and think things through.

THE SECOND KEY is to both look at the big picture and pay attention to the details. A broad perspective is essential to achieving a big vision. Taking the overall view of your organization prevents you from falling into sectionalism within your company, which may hinder the company's growth and eventually lead the whole organization to fall apart.

While always seeing the big picture, you also need to pay close attention to the individual tasks of your day-to-day work so as not to overlook small errors. Remember that the accumulation of small steps will eventually lead to success. Both a broad perspective and a detailed perspective are essential for executive managers.

THE THIRD KEY is to find and fix bottlenecks. As your organization develops, you will inevitably run into issues that hamper the company's further growth. But if you can discover and remove the bottlenecks, you can open up a path and progress to the next stage of development. You need to stay alert for any issues that prevent you from developing your organization. And once you spot them, make every effort to find solutions for them. As a top executive, you should also be aware of the risk that you yourself can become the bottleneck, so you need to keep cultivating and developing yourself.

THE FOURTH KEY is to always be mindful of customers' needs. Your customers—those who use your products or services—ultimately make the final assessment of your business. The decisions of top executives should always be made based on their customers' perspectives.

THE FIFTH KEY is to increase the added value of your products and services. Added value is gauged by the total sum of your customer satisfaction with three elements—the quality of your product, your distribution and delivery operation, and customer services, including sales and maintenance.

Regardless of whether your organization is for-profit or nonprofit, it has to satisfy its customers in these three areas to increase the added value of what you offer. In the case of Happy Science, increasing the added value of our products and services depends on the contents of my lectures and books, the plans and decisions of the executive staff in running the operations of our faith centers, and various activities that we take part in. Increasing the total added value of the organization is essential to gaining recognition in society.

I hope that these five keys to developing a management mentality will help you continue your efforts to become an effective and successful manager.

CHAPTER FOUR

Overcoming Recession

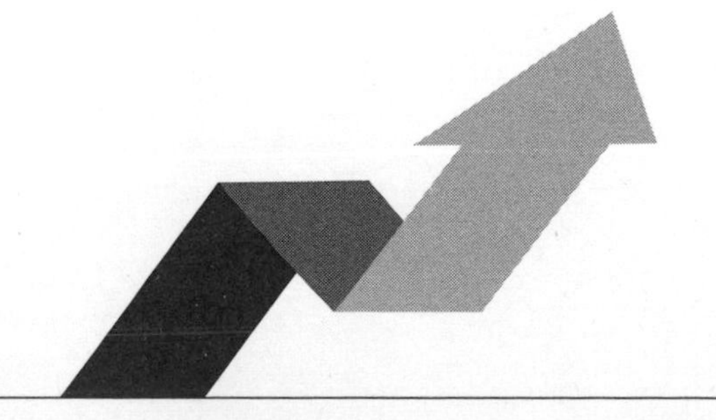

Six Useful Perspectives *during* Transition Periods

◀ ONE ▶

Finding the Causes of Recession

THE COLLAPSE OF THE Bubble Economy: The Example of Japan

It is probably uncommon and perhaps inappropriate for a religious figure such as myself to become too involved in complicated political matters, but since economic recession has a significant impact on the well-being of many people, I think that this is a subject worth looking at from a religious point of view.

Before discussing how we can counter recession, I would like to first show what triggers recession and how it may affect us using the example of an economic downturn that Japan has suffered since the beginning of the 1990s. This prolonged recession has caused a significant amount of anxiety among the Japanese people. Although the government has announced that the economy had bottomed out and begun to grow again, the general populace hasn't bought it. Indeed, the Japanese economy has continued to be slow since that announcement.

One of the main causes of the economic downturn was the press's all-out denouncement of the bubble economy, which pressured politicians and government officials to jump on the bandwagon to burst the bubble. This strong sentiment against the bubble economy, in my opinion, has its roots in the left-wing ideology that runs deeply in the minds of bureaucrats, politicians, and members of the mass media. The leftist idea that capitalists are evil became rampant among them after the end of the Cold War.

The press can serve the public interest when it discloses and condemns businesses that use illegal or unethical practices to make easy money. But its chorus of criticism against booming businesses in the 1990s significantly damaged the precious spirit of capitalism.

THE QUESTION:
Was It Creative Destruction?

I came to realize, over the years, that bursting the bubble was perhaps exactly the scenario the policy makers had planned back in 1991. The then prime minister Kiichi Miyazawa called for the price of housing to drop to the equivalent of five years' wages, and the trend indeed headed in that direction. Looking back, I feel that the government and the Ministry of Finance at that time knew full well what to expect in years to come, and that the recession may have been part of their plan from the start.

Upon reflecting on the series of events that led to the collapse of the bubble economy, I realized that there were basically only two ways to implement Miyazawa's policy of enabling people to buy a house in the Tokyo metropolitan area for a price equivalent to five years' income. Either people's incomes needed to substantially increase or housing prices or construction costs had to drop significantly.

At the beginning of the 1990s, Japan boasted one of the highest GDPs per capita in the world, and the average income was already high. So it would have been almost impossible to increase earnings manyfold. That meant that the only way to achieve Miyazawa's goal was to drop land prices. So I believe that the government could have foreseen the subsequent collapse of the financial institutions and the decline of the construction industry as a consequence of implementing this policy.

They must have known that they were heading for trouble, but the results were far-reaching. The administration collapsed, the ruling party lost its power, and politicians left their parties, realigned with new parties, and formed a new coalition government. The Ministry of Finance and other government offices came under severe scrutiny and disbanded, while the bureaucratic system collapsed. This was the extent of the reaction their policy caused.

The question we have to ask is whether we are under going a process of mere destruction or creative destruction that will lead to a new era.

The question we have to ask is whether we are undergoing a process of mere destruction or creative destruction that will lead to a new era, the coming age. Sheer destruction would only lead to the collapse of Japan. But if it is indeed creative destruction, as Joseph Schumpeter has argued, then we will have to bear the pain of parting with the old ways and prepare ourselves to build a new future.

Unfortunately, the leading politicians of Japan have yet to present a clear vision of the country in the coming century and beyond. They remain reactive to the issues at hand and are often too slow in responding to the situations that the country is facing.

TWO

Assessing the Societal Impact of Recession

THE IMPACT OF the Real Estate Revolution

In hindsight, we could have avoided the recession if the government had refrained from triggering the bubble burst in the early 1990s. But what's done is done, and there is no turning back. So we now have to think about what needs to be done and how we can cope with the prolonged recession.

The first factor we need to consider is that a real estate revolution and a sweeping reshuffle of the capitalist class occurred as a result of the recession.

With the end of World War II came the collapse of the old regime and the rise of a new breed of landlords and entrepreneurs who brought about the economic growth of the following fifty years. Those who started new businesses right after World War II were able to buy land and real estate at low prices. These people made a great fortune, not necessarily through hard work, but mostly because of the subsequent rise in land prices. This gave birth to a new breed of nouveau riche.

The rise of land prices prevented newcomers from purchasing real estate and forced them to rent offices to run their businesses. On top of this, Japanese banks hesitated to lend money without collateral and would only lend to people who already owned real estate. This made it extremely difficult for new entrepreneurs and developers to launch a new business.

But the crash in land prices caused by the recession significantly reduced the net worth of financiers who owned real estate, and many of them were forced to relinquish their assets. The drop in land prices also meant that banks could no longer offer land-collateral loans, at least not in the relatively simple way they had when land prices were high.

From this perspective, the recession may have had the merit of bringing about a sweeping reshuffle of the capitalist class. And now is the time for us to undergo another transformation and bring innovation to our society.

CHANGES WE MUST MAKE to Move Toward the Future

Seen in this light, the recession can present new opportunities for upcoming industrialists and entrepreneurs. But for these people to make use of the circumstance, I believe that the public needs to be ready to undergo significant changes on a national level.

We need to find ways to bring prosperity to both business and banks. In a mutually beneficial relationship, when the business that the bank finances grows, the bank grows too. With increasing amounts of the company's savings and loans, the bank's interest income will increase as well.

During a time of recession, many banks face financial difficulties and even bankruptcy. This bleak economic climate will compel them to restructure their financial system based on land-collateral loans. Banks will need to evaluate businesses based on their potential for success. And to do this, they will need to study how corporations are built, developed, and managed. Financial statements alone will not teach them the substance of the work the businesses are involved with or how they manage to make a profit. Banks will need to start learning how businesses operate and expand their stock of knowledge about new businesses and upcoming industries.

At the same time, entrepreneurs will also need to study hard and develop a keen sense of business and of future trends if they are to receive unsecured loans to cultivate new industries.

Businesses that have developed on their own will continue to grow, but businesses that have enjoyed and benefitted from the comfort and protection of the government will not be able to survive a severe economic downturn. For example, agribusinesses and fishery businesses that largely depend on government subsidies will need to cultivate entrepreneurship to start earning profits on their own to stay in business. They will have to either downsize their businesses or go under unless they grow out of the need for government protection and transform themselves.

Another area that will see a significant change is the real estate industry. Falling land prices will mean that those who previously could not afford to buy land now have the opportunity to acquire land capital to expand their businesses or start new ones. Existing property owners, on the other hand, face an increasing risk of bankruptcy, and they will need to innovate with a new entrepreneurial spirit to survive the difficult times.

As land prices fall, construction costs will drop, and the construction industry will likely go into a slump. A relaxation of regulations will prompt more foreign construction companies capable of designing and building good-quality buildings for low prices to make inroads into the domestic market.

At least in the case of Japan, these changes should be welcomed. Although Japan is one of the wealthiest countries in the world, the cityscape is cluttered with dingy buildings and dilapidated housing. Greater liquidity in the land and construction markets will allow a variety of new and modern construction, drastically improving the cityscape and bringing about urban innovation. During this process of innovation, individuals and parties who were benefiting from the previous market will probably face a severe management crisis. But this is the cost we may have to bear as a nation to ensure the coming of the new era and to build the nation's future.

◀ THREE ▶

Investing in Yourself

THE INDIVIDUAL'S
Worth in Business Will Become Clear

We've seen the possible impact of a recession on a societal level, but how can we as individuals cope with such volatile times? During an economic downturn, many companies—both small and large—go bankrupt or go out of business. This will inevitably lead to an increase in unemployment.

With a rise in unemployment, we will start seeing a social trend in which people change jobs more frequently. When this happens, we will start evaluating people based on not the type of job they have, but their skills and capabilities. Our worth as individuals will be assessed more accurately and objectively.

We will also become more conscious of how much money we are generating in a year. When we work for a large corporation, we can't tell how much profit each worker is generating. When the business is doing well, we feel as though we are contributing to the company even if our work is unprofitable. Government-protected businesses in particular often have inefficient employees and even executives who are virtually out of work within the company.

During difficult economic times, each individual's worth will be clearly assessed based on their ability. They will be paid according to their capability, competence, and performance.

CULTIVATE YOURSELF
During a Recession

To increase your value in the job market, you need to invest in developing your abilities. As a matter of fact, the best investment you can make during a recession is yourself.

When the economy is slow, the interest rates are so low that money kept in a savings account will bear virtually no interest, and investing in securities can be risky. Some people may even consider keeping their savings in their own homes out of a fear that banks will go bankrupt.

> In times of recession, self-investment is probably the best way you can use your money to earn a return.

In times of recession, self-investment is probably the best way you can use your money to earn a return. Investing in yourself will probably bring in much more profit than the interest you can earn from a savings account or security. You can increase the value of your original investment of $10,000, for example, if your self-investment helps you open a path to a new career, speeds up your promotion, or results in a higher-paying offer from another company.

In this way, you can prepare for the time when the economy eventually picks up by investing in yourself. You can perhaps go to school or take a course to acquire new knowledge or spend your money on travel or other hobbies. If you have a family, investing in your children's education can also be an extremely profitable investment in the long run.

Recessions are often protracted over a long period of time. The pervading sense of stagnation in society may make you lose enthusiasm for work or feel depressed. But only by surviving these tough times can we become the leading figures of the next age.

◂ FOUR ▸

Reassessing the Value of Your Family and Spirituality

REDISCOVER THE VALUE of Your Family and Your Spiritual Life

Another way to cope with the recession is to use this time as an opportunity to reflect on your life and reassess the value of your family life.

When business is good, you may be occupied with your work and socializing with colleagues and clients. But during a recession, you have to tighten your purse strings, which prevents you from spending money on social activities. You may miss your social life, but you can take advantage of this time by rediscovering the value of your family life. I am sure that looking after your family is a great innovation you can make in your life; by doing so, you will be able to find new values.

A time of recession also lets us shift our focus from an ostentatious and extravagant life to an inner spiritual life. We feel the vanity of pursuing material wealth and turn to spiritual values. We become introspective and start to reflect on our way of life, feeling that we should put the brakes on our avarice and rethink our lives. It is a time when we can appreciate the value of faith and the significance of a spiritual life. It is a time we can look within, sharpen our spiritual senses, and realize the true value of spirituality in our time. In fact, a recession often brings about a boom in religion, because many people join religious groups or take part in their activities.

This period can offer you a chance to learn important life lessons. In the same way that people realize the value of their lives when they suffer from illnesses, workaholics can appreciate the value of their time when a recession makes them stop and reflect on their busy lives. Contemplating your life and rediscovering the joys of family and inner spiritual values can indeed make for very precious moments in your life.

POLISH YOUR SOUL through Spiritual Practices

Cultivating your spirituality is actually a good investment you can make during a recession. By developing your spiritual awareness, you can add depth to your character and develop your inner qualities. By polishing your mind through spiritual practices, you can grow to become a person who can attract and influence a lot of people. Spiritual awareness is in fact an essential quality to develop to become a top business executive.

In addition to rediscovering family values and cultivating your inner self, you can invest in acquiring new knowledge and skills through education and study. You can find many materials of good value for low cost. For example, books are inexpensive but valuable. A book may only cost ten to twenty dollars, but the information it contains is often worth much more than the price. You don't have to spend a fortune on books, as there is only so much you can read, but the knowledge you can gain from the books you buy can be priceless. I would like to emphasize again the importance of cultivating yourself during a recession.

> Spiritual awareness is in fact an essential quality to develop to become a top business executive.

FIVE

Discovering the Potential of Your Future Growth

REVIEW YOUR Business Operations

In a world where more than seven billion people live, there will always be a demand for various products and services, which will give rise to an economy of some kind or another. But we still need to accept the reality that people will become much more selective in times of economic downturn.

How should then business executives and owners cope with recession? One way is to use it as an opportunity to reassess their business operations and discontinue products and services that do not meet people's needs or are of little use to customers. They will also need to deeply contemplate what people really want.

During a recession, people live frugally and only buy products that offer good value. So the products and services that stop selling are those that people bought not because they really liked or wanted them but only because they had extra cash to spend. To put it another way, many of the products and services that continue to do well regardless of the economic climate are truly valuable to customers.

So as executives and managers, we should do all we can to pursue and develop products and services that will continue to sell even during a recession.

The recession isn't to blame for a decline in sales, a decline in profit, or the bankruptcy of your company. Rather, it is testing whether you can transform your business into a company that is strong enough to survive difficult times. This is the time for you to identify your company's weaknesses and focus on enhancing its strengths.

The recession may be a good time to restructure your business as a whole. You may want to consider altering the overall workflow and the basic structure of the company, including its cost structure and income structure. In this sense, a recession can awaken you to what it is that you need to do.

ALWAYS BE ON THE LOOKOUT for New Business Information

To be a top executive or entrepreneur, you need to always look for the seeds of your future business—your business of tomorrow, next year, the next decade, or the next era—and grow these seeds once you find them. You may not always find them by reading books. A lot of times, you need to go out into society and discover them through experience. In any case, the first step is to discover the seeds that you can develop into the bread-and-butter products of your business in the future.

Future trends will not appear out of nowhere; their beginnings exist in the present. You can find hints and ideas about future trends around you, in what people say and think, and in the information you find in the papers, in magazines, or on television. People are not always aware of the seeds of the next business trends. So you need to keep your eyes open to find these seeds in the present. Always ask yourself what the seeds of the future may be and what the seeds of coming industries may be, and search for those seeds in the present.

Always ask yourself what the seeds of the future may be and what the seeds of coming industries may be, and search for those seeds in the present.

It is up to you to develop the ability to detect new trends. You can significantly increase your chances of finding new information by proactively seeking and collecting it 365 days a year. Instead of desultorily waiting, keep an eye out for the seeds of tomorrow's business while simultaneously committing to self-development.

Top executives, in particular, should be quick to sense the coming trends. Executives need to do more than busy themselves with paperwork, meetings with the same people, and other daily routines. They need to constantly think about how the future will unfold and how the world will change. Allowing yourself enough room to see things from a comprehensive perspective is essential to successfully managing your business.

◀ SIX ▶

Finding New Business Opportunities

To sum up what I've discussed in this chapter, we can see, based on what Japan has gone through since the 1990s, that a recession brings about a transformation that changes industrial structures at a national level, and that this transformation can take ten years, if not longer. During this period, we will experience all types of dramas, from tragic to jovial, but whatever may happen, we have to remain tough and survive. In the end, we can triumph as long as we constantly seek the seeds of the future and keep cultivating ourselves.

Taking a proactive attitude of adopting new values, discovering people's needs, and offering products and services that meet the market's needs are essential to staying in business in the coming age. Those who believe that their jobs are secure just because they are working for a major corporation will end up being left behind, one after another. The same is true of those who rely on governmental protection and believe that they can keep on living off other people's taxes; the time will come when they are left with nothing. Whether they are civil servants or people who live off subsidies, they will need to cultivate entrepreneurship to survive the economic downturn.

There are various measures the government can implement to tackle the recession, and I can think of quite a few myself. But I would rather not talk about the macroeconomic perspective here, because at the individual level, we can always look toward the future, discover the seeds of new opportunities, and plant and nurture those seeds.

A recession is a period of transition that provides opportunities for new businesses. Tomorrow's leading entrepreneurs and ingenious inventors will emerge from among the people who are going through difficult times and may seem confused about what to do. Eventually, new businesses will replace existing ones. New entrepreneurs will launch big projects, which may give rise to new giant conglomerates.

In the midst of this great transformation, we need to pay close attention to things that happen around us so as not to get tripped up by trivial matters. We should also maintain a broad perspective and keep on cultivating and developing ourselves. This is how we can overcome recession.

CHAPTER FIVE

Optimizing Your Leadership Capability

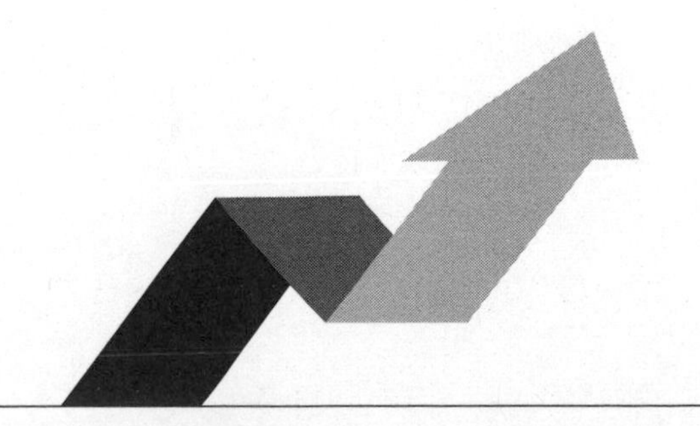

Six Pearls of Wisdom *for* Taking the Middle Way

◀ ONE ▶

Discovering the Pitfalls of Optimism

REACHING ONE'S LIMIT: The Example of a Japanese Prime Minister

In the spring of 2000, while in office, Japanese Prime Minister Keizo Obuchi died of a sudden stroke, which resulted in a change in leadership for the country. The main cause of the stroke was believed to be excessive stress he suffered from a couple of incidents that happened right around that time. One of them was the eruption of Mount Usu in Hokkaido in March 2000, and the other was the breakoff of his coalition government. It was also rumored among government officials that Prime Minister Obuchi had received divine retribution for causing a massive budget deficit. Others said that he had been punished because, despite his advocacy for a prosperous and virtuous country, he displayed a lack of virtue by joining forces with New Komeito Party, whose parent organization was seen by many as a misguided religion.

But as the saying goes, an outsider has the best perspective. Probably because I was looking at the situation from a distance, with no vested interest, I could clearly see what caused his death: he had simply reached the limits of his ability.

Although Obuchi had done much better than expected during his first term in office, his lack of ability and excessive lust for power became apparent in his second term, and he came under strict public scrutiny. He overextended himself, trying to win back popularity by hosting the world summit in Okinawa in 2000, but he couldn't handle everything and collapsed in the end.

It seems that he overworked himself because he wasn't confident of his ability to make policy decisions and create a vision for the nation's future. Instead of quietly contemplating national policy, he socialized with other political figures in an effort to compensate for his lack of ability. I am sure he felt chagrined at passing away like that, so I probably shouldn't be too critical of him. But from an objective perspective, it seemed that he had reached the limits of his ability and failed to retire from his post at the right time.

Japan has seen so many prime ministers come and go during the last few decades. I think this is because Japan's sudden growth and increasing influence on international affairs has become too much of a burden for one person to handle. Most of them exhausted themselves after one or two years because they didn't have the intellectual, theoretical, philosophical, or physical reservoirs that most American presidents who served a term of four years or longer did. But the fact that Japanese political leaders could not withstand a long-term administration shows that Japanese politics has rapidly become so complex and sophisticated that any one person's ability alone has not been able to bear the heavy burden of governing the entire nation.

IS THE GLASS
Half Full or Half Empty?

I bring up the topic of former Prime Minister Obuchi because his political career reminded me of a pitfall of optimism. His philosophy was exemplified by his answer to a common proverbial question, "Is the glass half full or half empty?" In a speech at the Diet, Obuchi said that he liked to take the optimistic view of looking at it as half full. This probably was his candid, personal view because it sounded too colloquial to be drafted by his bureaucrat speech writer.

This parable tests our individual perceptions. Whether you see a glass as half full or half empty determines how you see particular situations—as seeds of happiness or as seeds of unhappiness. This can certainly change your outlook on life. By focusing on what you lack, you will increasingly feel distressed and miserable, but by focusing on what you do have, you will feel blessed and contented. This idea is the origin and the starting point of optimism.

> Top leaders can make deadly mistakes if they overlook the facts and make decisions solely based on their subjective views.

Feeling grateful for what we have received can indeed benefit us spiritually. But there is one pitfall in this view. While we can find happiness by changing our perceptions, it doesn't change the fact that the glass is half empty.

Anyone who is responsible for leading a large group of people, whether a politician or a corporate executive, should never lose sight of the fact that our subjective views do not change the objective facts. Top leaders can make deadly mistakes if they overlook the facts and make decisions solely based on their subjective views.

A LEADERSHIP LESSON: Japan's Grave Mistakes during the Pacific War

On a personal level, optimistic thinking can benefit us significantly, but it can easily result in a disastrous outcome or invite evil when it comes to critical situations such as wars in which the lives of many people are at stake and business deals that hold the key to an entire organization's survival.

One example of how optimistic thinking can bring about disastrous outcomes is how the Japanese Army handled its southern operations during the Pacific War. The strategists at the Imperial Headquarters dispatched troops to the southern front, neglecting to offer logistic support, based on a false assumption that the troops would be able to find ample food in the jungles there. They simply looked at maps and decided that the troops would be able to procure food and water in the tropical jungles, without ever setting foot in the area themselves.

The troops had to go through hell. At Guadalcanal, an island in the southern Pacific, twenty thousand Japanese soldiers died from fighting, hunger, and disease. Another tragedy that struck them was at the Battle of Imphal, in which the Japanese Army attacked the British forces stationed in Imphal in an effort to assist the Indian National Army. The Japanese Army planned to cross the Arakan Yoma mountain range from Burma and enter India, but they had virtually no plan for procuring food, and the troops were ordered to advance by calling on their mental strength alone. After a devastating march, a total of approximately thirty thousand Japanese soldiers died. These tragedies could have been avoided if the commands had been given by someone who was well versed in the actual situation. But all too often, we build castles in the air and make wrong decisions based on our assumptions, without obtaining the necessary knowledge or experience.

It is generally believed that the main reason Japan lost to the United States in the Pacific War was its lack of resources. But if we look closely into the matter, it becomes clear that Japan lost largely due to a lack of information. The U.S. forces were fully equipped with radar and knew the positions of all the Japanese ships and aircrafts, but Japan had no idea where the enemy was. It was virtually impossible to win, because Japan had already lost the information war.

Japan had developed a rudimentary form of radar before the United States did but didn't put it to practical use, because Japan had not realized its importance. The United States, on the other hand, had developed radar technology for practical application, which gave them an advantage.

Japan had also proved the effectiveness of deploying naval aircraft carrier task forces during the Pearl Harbor attack, but it was the United States that learned and made the most out of the lesson.

Without realizing the gravity of deploying carrier-based task forces, the Japanese Navy kept concentrating its forces on battleships. On top of this, they kept losing battles because navy commanders held back from using the battleships for fear of losing them.

The phrase, "perspective can change everything" holds true in some situations. But we can't change objective facts—or make up the differences in technology and knowledge—simply by changing our perspective.

EASY SOLUTIONS
Can Invite Evil into Business and Life

How we see a glass of water probably won't have significant impact on our lives, but as we saw, dealing with food supplies for a large army is literally a life-or-death issue. How we look at the situation won't change the amount of food required for the troop to march for a certain period of days. Ignoring the reality will only make it a death march, no matter how positive or encouraging our perceptions may be. An objective assessment of the real circumstance is not a sign of cowardice or negative thinking; it is simply wisdom.

Even heavenly thoughts can't always prevent evil deeds. For example, no matter how hard we pray for a miracle, we can't expect divine intervention to prevent a shabby building from collapsing during an earthquake. Divine beings are not going to perform miracles only to save a building with unfilled spaces in its concrete blocks or a building that lacks reinforcing steel bars. They will not extend their helping hands that far. It is up to us to construct earthquake-proof buildings and expressways with a certain level of resistance. Even when we know that our spiritual devastation will call down divine wrath and bring about a natural disaster, we shouldn't just wait for it to come and do nothing; we should do what we can to keep the damage to a minimum.

Another example of someone who seems to have fallen into the pitfall of optimism is Masaharu Taniguchi (1893–1985), founder of a Japanese religious organization, Seicho-no-ie. In his book, *The Truth of Life* (although it may not be fair to criticize his book now, considering that it was written before World War II), he says that we should not think about illness at all, because merely thinking about illness will attract it. I agree that worrying ourselves sick about becoming ill will no doubt prevent us from working fruitfully. But then he goes on to say that we should never carry medicine with us when we go on trips because this may attract illness.

I don't think bringing medicine on your trip should be considered negative thinking that will invite evil; it is wisdom. When we go on a trip, we can't go to our home doctor, and we may not be able to find the type of medicine we need or appropriate medical facilities nearby, especially in the case of overseas travel. We carry medicine with us not because we want to become sick but as a precaution against illness.

Although leaving medicine behind while traveling so as to avoid attracting illness may be appropriate advice for some, it is probably not appropriate for the majority of people. It could well result in tragedy if a head of state tells the whole population not to take any medicine with them when they travel. This advice simply lacks wisdom.

If the optimistic perspective of looking at the glass half full were applied to governmental affairs, we might simply look at our national assets and conclude that our country is doing well financially, when in fact we have a large national deficit. This is nothing more than an irresponsible attitude.

Although optimistic thinking can bring about our personal growth on the individual level, without an objective perspective, it could bring evil when it comes to organizational, societal, and national matters.

◂ TWO ▸

Harnessing the Power of Wisdom

HUMBLY ASSESS
Your Own Capability

If there are particular abilities that you don't have yourself, surround yourself with great people, and you can still do a good job. This was the philosophy that steel magnate Andrew Carnegie put into practice. Carnegie even wished to have his tombstone inscribed with a similar message.

Attributing your success to the capable people that surround you instead of your own ability is certainly a humble attitude. But it would be a mistake to interpret his words to mean that anyone can use capable people. We have to take into consideration that in Carnegie's day, many capable people could not receive a formal education, simply because the education system was not yet well established. Andrew Carnegie received little formal education not because he was incompetent, but because he didn't have the opportunity.

It is not easy to manage people. We need deep wisdom, relentless effort, and a good system to make the most out of capable people. As I said earlier, optimistic thinking, if used as a principle of leadership, can lead to excessive dependence on others or abandonment of our own responsibilities.

Leaders need to be strict with themselves. They need to correctly assess their own abilities, their accumulated knowledge and efforts, and their comprehensive capacity, as well as the level of support they can expect to receive from others, the opportunities heaven is giving them, and the advantages they have over others.

NARROW DOWN YOUR IDEAS as Your Organization Grows

During the last several decades, I have been operating a variety of activities at Happy Science. At the onset, when we started small, nearly every idea we came up with became a success. Almost all the projects and events we launched turned out well while the organization was still small.

But once our organization grew to a certain size, we often found a way to further growth when we narrowed down our ideas and carried out only the ones we felt confident would be successful. We decided to drop some projects that required additional development and ideas that could have a negative impact on our organization as a whole.

As I have learned from my own experience, implementing every idea and project as it comes up may work for small organizations, but it doesn't work for midsize to large organizations. We have to polish, elaborate, and develop our plans until we feel confident of our success.

Unpolished ideas that we don't give ample thought and consideration to often end in failure.

When our organizations reach a certain size, we need to develop patience and restraint to relinquish some of our plans, taking into consideration the chances of success, the weight of our responsibilities, and the possible impact of our decisions on the entire organization. Sometimes, we should postpone projects until we find a capable leader who can carry them out successfully. We will rarely fail when we can carefully select and develop ideas this way.

Organizations continue to transform as they grow, so as leaders, we need to change the way we think accordingly.

Organizations continue to transform as they grow, so as leaders, we need to change the way we think accordingly. A very good idea that brought us success this year may not work five or ten years from now.

USE WISDOM FOR Both Offense and Defense in Business

Even a philosophy that has continued to help people solve their inner issues for thousands of years still needs to be carefully thought through before it is put into practice, particularly when it involves a large number of people, has significant impact on society, or requires interaction with other organizations and groups.

For instance, let's say you have a certain annual budget for the salaries of staff members at a religious organization you manage. To stay within the budget, you may take an optimistic view and say that even beggars don't starve nowadays, so staff members should certainly be able to get by with half of what they receive now. But this is not the correct way to deal with the issue. You need to do what you can to help the organization generate sufficient income so you can pay staff members appropriate salaries.

Similarly, you might think of cutting down on expenditures by negotiating with your landlord to lower your rent. If it's only for a short period of time, the landlord may accede to your request, saying, "All right, I understand that your organization is working hard for a good cause, so I will do my bit and waive your rent for six months." You could then share this story with others as a success story of optimistic thinking. But this example shares the same problem as the Japanese army's miscalculation: a lack of wisdom.

Pragmatic thinking is not necessarily the same as negative thinking. Seeing reality as it really is is different from believing in a bleak future. If anything, naïve optimism can bring about negative outcomes when applied to organizations; it can cause business stagnation, serve as a justification for inaction, or even become sophistry. To avoid falling into this pitfall, we need to use the power of wisdom. The power of wisdom will strengthen not only defense but also offense and let us continue winning in management.

◀ THREE ▶

Thwarting Evil Intentions

CAREFULLY EXAMINE
Your Capabilities

Although we have complete freedom over our minds, we cannot become entirely different people. This is because we each are born with different innate abilities, such as physical strength, intelligence, and aptitude. We can only change some aspects of our personalities, which are, to a great extent, what define our being.

So we should objectively assess our characters, capabilities, and physical strength, in conjunction with our work environments and relationships with family and friends, and consider how much we can achieve, instead of merely wishing that we will get lucky in the end.

Your goal should not be a pure fantasy or an abstract idea. Ponder your ideal self, or the best you can be, based on your potential. By calmly assessing your abilities and the surrounding circumstances, you will be able to see how you can get the most out of yourself.

When examining yourself, consider your relationships with others as you try to realize the vision in your mind. Check whether the achievement of your goal may cause any harm to people around you. The image of your ideal self should not be excessively egoistic to the point that it hurts others or brings them unhappiness.

THE PSYCHIC WAR between Hitler and Churchill

One of the laws of the mind states that we become what we think. We become the kinds of people we believe ourselves to be. Things often turn out the way we engrave them on our minds because the thought patterns that we hold in our minds take shape and manifest themselves in reality.

This same law applies not only to ourselves but to others as well. So a vision may fail to manifest itself when someone has a strong will against it. One example of this is the psychic war between German politician Adolf Hitler and British statesman Winston Churchill. Hitler must have had a vision of the Nazi domination of Europe and the creation of a European empire under his rule. This vision came true to a certain extent when he was at the brink of victory.

But the fruition of Hitler's vision was blocked by Winston Churchill's opposing vision. Believing that the Nazis were evil, he held a strong vision of completely defeating them. In this way, Hitler's vision of European domination and Churchill's vision of destroying Hitler and fascism unfolded and clashed with each other. Their powerful visions captured the hearts of many, leading them to war.

From a spiritual perspective, it was a psychic battle or a duel between two sorcerers. Hitler and Churchill—two wizards with enormous power to control millions of people—each had a vision that clashed with the other's vision, and they launched a psychic war. When the two parties' power is equally strong, it is rare for one party to achieve total victory. One party has to have about ten times as much power as the other to win a complete victory, but this usually isn't the case.

It is often believed that our visions will come to fruition whether they are good or bad. And this certainly seems to be true on an individual basis. But in reality, all our thoughts are subject to the interposition of right and wrong to some degree.

> In reality, all our thoughts are subject to the interposition of right and wrong to some degree.

If you so desire, it is quite possible for you to go out and commit a crime today. You could attack someone unawares and hit him with a club. Most people—even fencing experts—would not be able to dodge such a sudden move. But you would most likely be caught right away and restrained from assaulting more people on the street. In this way, an opposing power will work to prevent someone's vision from coming true.

No one can keep committing crimes forever, because someone with the vision of capturing the criminal will eventually stop their criminal activities. Thieves and robbers may realize the visions of committing burglary and theft. But at the same time, the police will start chasing after them with the vision of arresting them. Even the most ambitious people can rarely fulfill their visions completely, because there are always opposing forces, or those who hold the opposite visions.

USING THE POWER OF THOUGHT as a Deterrent Against Evil

Looking back on history, some people succeeded in fulfilling their evil ambitions, so malicious intentions are not necessarily doomed to fail. But their realization can be blocked if someone can see through their ulterior motives.

An example of this is how the guru of the Aum Shinrikyo, a Japanese doomsday cult, failed to realize his horrific ambition of conquering and ruling Japan. He held a grudge against society because he felt that society had oppressed and rejected him in the same way that his parents had. He entrusted Shiva, the Indian god of destruction, and tried to justify his vengeance by claiming to be a reincarnation of Shiva. He sought revenge on society and developed a desire to subjugate the country.

Beginning in about 1991, I could clearly see the evil ambition of Aum Shinrikyo, and the organization's involvement in criminal acts. So I publicly criticized them for their wrongdoings. I made this clear in an interview with the press, so some members of the media were already aware that I was accusing Aum Shinrikyo of their misconduct. In the end, his vision did not come to fruition because I was able to see through his ulterior motive and had the power to counter it.

As we have seen, we can realize our vision to a certain degree, but when someone develops an opposing vision, the two visions will clash and may cancel each other out, so that one vision reduces the impact of the other or causes it to lose out completely. So our wishes may not always come true, and it often has to do with whether our ambitions are good or evil. Many people are determined to prevent evil acts, and their thoughts serve as deterrents to evil acts. So those who harbor evil thoughts will sooner or later find that their desires fail to come to fruition.

◂ FOUR ▸

Aiming for the Best in Managing Your Business

A LACK OF WISDOM CAN Cause the Fall of a Business: The Example of a Japanese Supermarket

If bad thoughts are often prevented from fully manifesting themselves, what about good thoughts or wishes to achieve success and prosperity? Let's say that you own a shop and you are wishing for your business to thrive, which would apparently be a good thing for you. But your business may stop growing if you go too far with your efforts to realize your wish.

When you wish to expand your business, consider whether the realization of your wish will bring about a good outcome on the whole. If, for example, your shop sells bad products at high prices and you try to expand your business by driving other small and honest owners out of business, your success will actually cause more harm than good. This will probably prevent your business's further growth.

Another element that may curb your business development is a lack of wisdom. An example of this is how one Japanese major supermarket chain expanded its business overseas but eventually went bankrupt.

The rise and fall of this supermarket was depicted in a popular Japanese television drama called, *Oshin*. The main theme of this drama was that life is a bundle of good and evil; no matter how hard things may seem, we can improve our lives through ardent effort, but if we allow ourselves to become conceited and live off our favorable circumstances, we will fall into poverty again. The main character of this drama goes through this cycle of fortunes and misfortunes repeatedly. In one episode of this drama, the main character's supermarket goes bankrupt after her son takes over its management, and the son gets indicted for window dressing. It was as if this drama predicted the future of this major supermarket chain.

What went wrong with the management of this supermarket chain? The supermarket was able to achieve a certain level of success, but what the owner probably didn't fully realize was that its growth was greatly helped by the economic boom of the postwar years; it was only in the natural course of events that a small retail store grew to a big supermarket. When other supermarkets started emerging, however, it faced market competition, which put its products and its managers' skills to the test. This prevented the supermarket from achieving its owners' vision of further business expansion.

The same thing happened to many Japanese electronics companies after World War II. The postwar development of the Japanese economy helped almost all the electronic companies grow, but once they all expanded their businesses to a certain size, they started coming into conflict with one another.

The IT industry grew rapidly as people became aware of the

many ways that computers could enhance their lives. (Before that, many people had believed that computers could only be used for complex calculations.) The computer industry's growth made existing IT companies feel as though they could keep developing forever, but this was not the case. Many new companies that offered similar products and services started emerging, triggering competition. As a result, the weaker companies were forced out of business.

As these examples illustrate, your wishes will come true to a certain extent. But at the same time, you need to know that things may go well during good times but that you also face a risk of big failures, such as bankruptcies, when the going gets tough. This wisdom is essential for your continued business success.

A SELF-INDULGENT OPTIMISM Invites Business Failure

Let me go back to the example of the Japanese supermarket chain to illustrate another important factor that may trigger a business's failure. Behind the supermarket's initial success was the owner's optimistic belief in its future. (The owner adopted the philosophy of a New Thought Japanese religion, Seicho-no-Ie ["house of growth"], in its management.)

Although the supermarket achieved a certain level of success in the rural areas of Japan, it couldn't open stores in Tokyo. Only the best of the best in Japan can survive in Tokyo. And a business success in Tokyo can often be a good indication of its success overseas. In other words, without winning in the Tokyo market, it would be highly unlikely, if not impossible, for a business to win in the global market. It may still be able to go into the overseas market, but it will not be able to win in competition against businesses that succeeded in Tokyo.

This was exactly what happened to this supermarket chain. It failed in its first overseas venture in Brazil, but it didn't learn its lesson. Instead, it went on to expand its business in Asia and failed yet again. The owner probably thought that the supermarket had a chance to succeed in developing Asian cities, which he thought resembled the provincial towns of Japan. When he opened stores overseas, however, customers demanded better products, and the supermarket lost out in the competition against other major grocery store chains.

The owner of this supermarket chain seemed to have been optimistic about its business growth in China. He probably saw a vast potential market there simply because its population size was ten times bigger than that of Japan. But this doesn't mean that the Chinese market has the same purchasing power as the Japanese market does.

I once watched a documentary about a hardworking Chinese worker whose salary was about two hundred dollars per year, which is less than one-hundredth of the average annual income in Japan. It will take many years before the average Chinese income equals the average Japanese income. The supermarket chain made heavy investments in China without taking this into consideration. Sad to say, it was only natural that it went bankrupt.

What invited this failure was a lack of wisdom; the owners misinterpreted the situation with an optimistic outlook. Focusing only on growth and expansion can make us optimistic, but if we go too far, we could end up going bankrupt instead of growing our businesses. We need to think carefully and deeply about the possible outcome.

If I had been managing the supermarket, I would have opposed the idea of expanding overseas immediately after succeeding in rural cities in Japan. I would have said that the supermarket needed to achieve success in Tokyo first.

In this day and age, information reaches every corner of the world instantaneously. Even in secluded areas, we can't expect people to remain ignorant of the better products and services that are available in the market. We need to be able to provide what customers demand, whether at home or abroad, which are good products and services.

ASSESS YOUR BUSINESS from Both Subjective and Objective Perspectives

Generally speaking, if we think good thoughts, good things will come. If we wish for growth, our businesses will grow. But we should always keep in mind that we coexist with other people who are also searching for prosperity in their lives.

Whether as an individual or as a leader of an organization, we each need to have both an objective perspective and a subjective perspective. For instance, you may insist that your company offers the best products available in the market, but you need to check whether that is true from an objective perspective, or from the perspective of the customer who is looking for a certain product. Of course, you can be confident of the products and services you offer, if that is an objective fact. But you need to also remember that all the other stores are also striving to offer the best products and services.

Objectively assessing your capability is essential to determining the borderline between appropriate growth and selfish desire. Even if your initial intention for starting your business was a good one, it can end up becoming a self-serving wish if it becomes more than you can handle. Knowing this is crucial to your business success.

> Objectively assessing your capability is essential to determining the borderline between appropriate growth and selfish desire.

The optimistic view of looking at the glass as half full can certainly make us feel uplifted, while the pessimistic view of looking at it as half empty may make us feel down. But it becomes an entirely different story if we see this glass of water from the perspective of the many people who want to drink water. Half a glass may be enough to quench your thirst alone, but if many others are thirsty, you need to calculate how much water you'll need for them. If you are to give one glass of water to each person, you can figure out the amount of water you need by counting the number of people who are thirsty. This is where your insight and wisdom as a leader is put to the test. You need to see through the situation and arrange for the correct amount of water. Forcing your way through with an optimistic outlook is not the solution.

While political leaders are expected to have the ability to handle national issues, company executives are expected to have the same type of ability to handle management issues. In most cases, it takes one person to start a business. And in the cases of small to mid-size companies, whether or not the business becomes a success depends on the owner's ability, talent, creativity, and ideas more than 90 percent of the time. But it is also the limits of the owner's ability that will lead the company to bankruptcy. That's why, once your company has reached a certain size, you should learn your limitations and determine whether your assistants can take over some of your work. But even with the help of your staff, there may be some tasks, situations, or responsibilities that your company cannot handle, so it's crucial that you determine the limitations of your business.

Another factor that determines your business success is whether your business's industry is on the rise. During an economic boom, most owners can grow their businesses, because there is usually a greater demand for their products and services. But only truly good products and strong companies can survive times of recession.

◂ FIVE ▸

Knowing the Limits of Your Abilities

A RECESSION PREVENTS BUSINESSES from Depending on Outside Funds

---

When the economy enters a long recession, many businesses are driven into bankruptcy. Even large corporations and government ministries are forced to downsize, which leads to increase in unemployment. This indeed becomes a serious issue to the people involved, but we can also find a positive aspect to it.

Downsizing and restructuring let the organization cut down on any unnecessary operations and excess personnel. Hard economic times inhibit businesses from steadily increasing their revenue and so keep them from continuing to tolerate employees who receive a promotion and an automatic raise every year without having contributed to the company.

Many banks start going bankrupt, too, and this often opens our eyes to the shocking reality that they are actually debt-financed businesses. They do business using deposits collected from their customers; in other words, they can only operate by borrowing money from the people.

Major banks have loans of hundreds of billions of dollars. If they have deposits of 300 billion dollars, for instance, that means that they owe 300 billion dollars and also have to pay interest on those loans. If the economy grows and people's deposits increase, banks will be able to earn interest based on the additional deposits and pay back the loans. In fact, the Japanese banking sector grew after World War II by repeating this process.

Japanese trading firms also relied heavily on a large number of loans to run their businesses. They did not repay the loans; all they did was to revolve the funds so that they could earn sufficient money on this capital to pay their salaries to their employees and interest to the banks. But the recession prohibited them from operating their businesses this way.

The Japanese banks found themselves in the same predicament once the economy stopped growing. The banks could no longer pay the interest on their loans, and the interest rate for savings deposits fell to virtually zero percent. This shows that the Japanese banks were simply revolving the deposited money and that they were not adding any value to their services.

Banking businesses can add value to their services by offering financing to small and unknown startup businesses to help them grow. If, as a result, the businesses that receive loans from the banks thrive, the banks will have produced added value for the money they were entrusted with. Whether a bank can help produce and grow businesses that start out with nothing can be the measure of the bank's true capability. But unfortunately, Japanese banks failed to develop such competence.

When the economy slows down, institutional failings come to light. Whether the institution is a corporation, a bank, or a governmental office, its operations management comes under scrutiny and undergoes reexamination. And this circumstance requires institutional leaders to acquire true wisdom.

DEBTOR NATIONS
Will Face a Crisis

It is difficult to lead a decent life with a crushing burden of debt. We may be able to get by for a while, but if we keep borrowing money from loan sharks, the day will come when we go bankrupt. We may be forced to leave everything behind and skip town or, in the worst case, feel desperate and even consider ending our own lives. In the same way, businesses will also go down if they accumulate a massive number of loans that they cannot repay.

The same principle applies to running a country; a nation deep in debt will stop functioning and may collapse. The government of a debtor nation should be examined based on a guideline of whether it is doing a job that's worth the amount of money it owes. If the administration is only concerned about the economic outlook for the next few years, then it needs to fundamentally change the way it runs the country.

In the United States, the Clinton administration reportedly improved the economy for almost ten consecutive years and reduced the fiscal deficit, which was one of the twin deficits, but this didn't mean that the country became debt-free.

What the Clinton administration did was raise interest rates to attract investments from overseas, particularly from Japan, and used this borrowed money to induce consumption to boost the economy. But this didn't change the fact that the United States was still deeply in debt. It was running an account deficit, and the American people were spending money that they hadn't earned.

Figuratively speaking, the United States at that time was like someone who enjoys gourmet meals and feels as if he's becoming stronger, but then has to fight obesity and treat weakening internal organs. Just as we as individuals cannot continue living above our means, countries have to find a way to live within their means. Otherwise, they will face a looming crisis in the future. A country that's heavily dependent on loans from other countries will inevitably face hard times.

◀ SIX ▶

Using Wisdom to Take the Middle Way

Optimizing our capabilities is essential for achieving continuous success not only on an individual basis, but also on organizational and national bases. We have to always seek our ideal state, which is not about garnishing our life with vain wants and extravagance.

Our optimal state of being is not an extreme state. It's neither about being too easy nor too strict with ourselves. It is not about bringing only benefits or only harm to our company. It is not about declaring such extreme statements as "illness does not exist," nor "illness is fate."

We should be aware of the limits of our capabilities, because excessive desires can destroy us in the end. But so can a lack of desire. Both greed and a lack of the drive to live will bring us ruin. This is an important principle to keep in mind.

> Through the lens of wisdom, look closely at yourself and the capability of the organization, and cultivate the strength to survive in an ever-changing environment.

Wisdom exists in between extreme perspectives. We should use the power of wisdom to find and take the middle way. Through the lens of wisdom, look closely at yourself and carefully assess the capability of the organization you belong to, and cultivate the strength to survive in an ever-changing environment.

Being aware of the boundaries of our thoughts is essential, because they will eventually manifest as outcomes. This principle applies to everyone, from individual citizens to the nation's top leaders. I hope that this chapter will help you optimize your leadership capability.

Afterword

In this book, I have presented more than just a theory of leadership success. I have offered basic principles of all-around success in life, followed by a concrete discussion of winning leadership strategies, keys to successful business management, ways to overcome a recession, and the wisdom of taking the middle way in politics and economics.

Although I, Ryuho Okawa, am mostly known as a religious figure, the ideas in this book highlight my role as a strategist, as well as the nation's leader whose purpose is to liberate the people living today from worldly sufferings. This book also offers a glimpse of my stern side as the leader and commander of the invincible organization Happy Science who does not yield an inch against the forces of evil.

This compilation of the numerous management tactics that I have personally tested and used in the real world for the last several decades will surely become a beacon of triumph for many people.

May the wind of Great Compassion blow throughout the entire world.

Ryuho Okawa
Founder and CEO
Happy Science Group

About the Author

RYUHO OKAWA is a global visionary, renowned spiritual leader, and internationally best-selling author with a simple goal: to help people find true happiness and create a better world.

His deep compassion and sense of responsibility for the happiness of each individual has prompted him to publish over 2,300 titles of religious, spiritual, and self-development teachings, covering a broad range of topics including how our thoughts influence reality, the nature of love, and the path to enlightenment. He also writes on the topics of management and economy, as well as the relationship between religion and politics in the global context. To date, Okawa's books have sold over 100 million copies worldwide and been translated into 29 languages.

Okawa was born in 1956 in Tokushima, Japan. After graduating from the University of Tokyo with a law degree, he studied international finance at the Graduate Center of the City University of New York and worked at a major Tokyo-based trading firm. In 1986, he renounced his successful business career to found Happy Science as a spiritual movement dedicated to bringing greater happiness to humankind by uniting religions and cultures to live in harmony. Happy Science has grown rapidly from its beginnings in Japan to a worldwide organization with over twelve million members. Okawa is compassionately committed to the spiritual growth of others. In addition to writing and publishing books, he continues to give lectures around the world.

About Happy Science

Happy Science is a global movement that empowers individuals to find purpose and spiritual happiness and to share that happiness with their families, societies, and the world. With more than twelve million members around the world, Happy Science aims to increase awareness of spiritual truths and expand our capacity for love, compassion, and joy so that together we can create the kind of world we all wish to live in.

Activities at Happy Science are based on the Principles of Happiness (Love, Wisdom, Self-Reflection, and Progress). These principles embrace worldwide philosophies and beliefs, transcending boundaries of culture and religions.

The Principles of Happiness

LOVE teaches us to give ourselves freely without expecting anything in return; it encompasses giving, nurturing, and forgiveness.

WISDOM leads us to the insights of spiritual truths, and opens us to the true meaning of life and the will of God (the universe, the highest power, Buddha).

SELF-REFLECTION brings a mindful, nonjudgmental lens to our thoughts and actions to help us find our truest selves—the essence of our souls—and deepen our connection to the highest power. It helps us attain a clean and peaceful mind and leads us to the right life path.

PROGRESS emphasizes the positive, dynamic aspects of our spiritual growth—actions we can take to manifest and spread happiness around the world. It's a path that not only expands our soul growth, but also furthers the collective potential of the world we live in.

Programs and Events

The doors of Happy Science are open to all. We offer a variety of programs and events, including self-exploration and self-growth programs, spiritual seminars, meditation and contemplation sessions, study groups, and book events.

Our programs are designed to:

- Deepen your understanding of your purpose and meaning in life
- Improve your relationships and increase your capacity to love unconditionally
- Attain a peace of mind, decrease anxiety and stress, and feel positive
- Gain deeper insights and broader perspectives on the world
- Learn how to overcome life's challenges

... and much more.

For more information, visit happyscience-na.org or happy-science.org.

“Hints for Success” Special Seminar

Master the 10 ultimate rules and become a person of great success!

This seminar will help you realize success in your life through the power of your thoughts and the power of your mind, and you will acquire the ways of thinking that you will need to achieve success in every situation in your life. The 10 practical rules are what Master Okawa himself practiced when he established Happy Science. By repeatedly putting these rules into practice, you will surely achieve larger results than you can ever imagine.

3 RITUAL PRAYERS recommended for this special seminar

PRAYER FOR FUTURE SUCCESS
PRAYER FOR THE GROWTH OF ANGELS OF WEALTH
PLEIADES PRAYER TO TURN THE WHEEL OF FORTUNE

Please contact your nearest Happy Science center for more information.

International Seminars

Each year, friends from all over the world join our international seminars, held at our faith centers in Japan. Different programs are offered each year and cover a wide variety of topics, including improving relationships, practicing the Eightfold Path to enlightenment, and loving yourself, to name just a few.

Happy Science Monthly

Our monthly publication covers the latest featured lectures, members' life-changing experiences and other news from members around the world, book reviews, and many other topics. Downloadable PDF files are available at happyscience-na.org. Copies and back issues in Portuguese, Chinese, and other languages are available upon request. For more information, contact us at tokyo@happy-science.org.

Contact Information

Happy Science is a worldwide organization with faith centers around the globe. For a comprehensive list of centers, visit the worldwide directory at happy-science.org or happyscience-na.org. The following are some of the many Happy Science locations:

United States and Canada

New York
79 Franklin Street
New York, NY 10013
Phone: 212-343-7972
Fax: 212-343-7973
Email: ny@happy-science.org
Website: newyork.happyscience-na.org

Los Angeles
1590 E. Del Mar Blvd.
Pasadena, CA 91106
Phone: 626-395-7775
Fax: 626-395-7776
Email: la@happy-science.org
Website: losangeles.happyscience-na.org

San Francisco
525 Clinton Street
Redwood City, CA 94062
Phone&Fax: 650-363-2777
Email: sf@happy-science.org
Website: sanfrancisco.happyscience-na.org

Orange County
10231 Slater Ave #204
Fountain Valley, CA 92708
Phone: 714-745-1140
Email: oc@happy-science.org

Florida
5208 8th St.
Zephyrhills, FL 33542
Phone: 813-715-0000
Fax: 813-715-0010
Email: florida@happy-science.org
Website: florida.happyscience-na.org

San Diego
7841 Balboa Ave. Suite #202
San Diego, CA 92111
Phone: 619-381-7615
Fax: 626-395-7776
E-mail: sandiego@happy-science.org
Website: happyscience-la.org

Atlanta
1874 Piedmont Ave. NE
Suite 360-C
Atlanta, GA 30324
Phone: 404-892-7770
Email: atlanta@happy-science.org
Website: atlanta.happyscience-na.org

New Jersey
725 River Rd. #102B
Edgewater, NJ 07020
Phone: 201-313-0127
Fax: 201-313-0120
Email: nj@happy-science.org
Website: newjersey.happyscience-na.org

International

Hawaii
1221 Kapiolani Blvd. Suite 920
Honolulu, HI 96814
Phone: 808-591-9772
Fax: 808-591-9776
Email: hi@happy-science.org
Website: hawaii.happyscience-na.org

Kauai
4504 Kukui Street
Dragon Building
Suite 21 Kapaa, HI 96746
Phone: 808-822-7007
Fax: 808-822-6007
Email: kauai-hi@happy-science.org
Website: kauai.happyscience-na.org

Toronto
845 the Queensway Etobicoke,
ON M8Z 1N6, Toronto Canada
Phone: 1-416-901-3747
Email: toronto@happy-science.org
Website: happy-science.ca

Vancouver
#212-2609 East 49th Avenue
Vancouver, BC,V5S 1J9 Canada
Phone: 1-604-437-7735
Fax: 1-604-437-7764
Email: vancouver@happy-science.org
Website: happy-science.ca

Tokyo
1-6-7 Togoshi,
Shinagawa Tokyo,
142-0041 Japan
Phone: 81-3-6384-5770
Fax: 81-3-6384-5776
Email: tokyo@happy-science.org
Website: happy-science.org

London
3 Margaret Street, London,
W1W 8RE
United Kingdom
Phone: 44-20-7323-9255
Fax: 44-20-7323-9344
Email: eu@happy-science.org
Website: happyscience-uk.org

Sydney
516 Pacific Hwy Lane Cove North,
NSW 2066 Australia
Phone: 61-2-9411-2877
Fax: 61-2-9411-2822
Email: sydney@happy-science.org

Brazil Headquarters
Rua. Domingos de Morais 1154,
Vila Mariana, Sao Paulo,
SP-CEP 04009-002 Brazil
Phone: 55-11-5088-3800
Fax: 55-11-5088-3806
Email: sp@happy-science.org
Website: cienciadafelicidade.com.br

Jundiai
Rua Congo, 447, Jd. Bonfiglioli
Jundiai-CEP 13207-340 Brazil
Phone: 55-11-4587-5952
Email: jundiai@happy-sciece.org

Seoul

74, Sadang-ro 27-gil,
Dongjak-gu, Seoul, Korea
Phone: 82-2-3478-8777
Fax: 82-2-3478-9777
Email: korea@happy-science.org
Website: happyscience-korea.org

Taipei

No. 89, Lane 155,
Dunhua N. Road, Songshan District,
Taipei City, 105 Taiwan
Phone: 886-2-2719-9377
Fax: 886-2-2719-5570
Email: taiwan@happy-science.org
Website: happyscience-tw.org

Malaysia

No 22A, Block2, Jalil Link
Jalan Jalil Jaya 2, Bukit Jalil 57000
Kuala Lumpur, Malaysia
Phone: 60-3-8998-7877
Fax: 60-3-8998-7977
Email: malaysia@happy-science.org
Website: happyscience.org.my

Nepal

Kathmandu Metropolitan City Ward
No.15, Ring Road, Kimdol, Sitapaila
Kathmandu, Nepal
Phone: 97-714-272931
Email: nepal@happy-science.org

Uganda

Plot 877 Rubaga Road
Kampala P.O. Box 34130
Kampala, Uganda
Phone: 256-79-3238-002
Email: uganda@happy-science.org
Website: happyscience-uganda.org

About IRH Press USA

IRH Press USA Inc. was founded in 2013 as an affiliated firm of IRH Press Co., Ltd. Based in New York, the press publishes books in various categories including spirituality, religion, and self-improvement and publishes books by Ryuho Okawa, the author of 100 million books sold worldwide. For more information, visit OkawaBooks.com.

Follow us on:

FACEBOOK: OkawaBooks

TWITTER: OkawaBooks

GOODREADS: RyuhoOkawa

INSTAGRAM: OkawaBooks

PINTEREST: OkawaBooks

Books by Ryuho Okawa

Invincible Thinking

An Essential Guide for a Lifetime of Growth, Success, and Triumph

Hardcover | 208 pages | $16.95 | ISBN: 978-1-942125-25-9

Invincible Thinking is the dynamite that lets us open a crack of possibility in a mountain of difficulties, the powerful drill that lets us tunnel through the solid rock of complacency and defeatism and move steadily ahead toward triumph. A mindset of invincibility is your most powerful inner tool for transforming any event or circumstance into inner wisdom and soul growth. Invincible thinking will give you all the nourishment you'll ever need to fulfill your purpose in life and become a guiding light for others.

The Laws of Success

A Spiritual Guide to Turning Your Hopes Into Reality

Softcover | 208 pages | $15.95 | ISBN: 978-1-942125-15-0

The Laws of Success is the modern world's universal guide to happiness and success in all aspects of life. You will find timeless wisdom, the secrets of living with purpose, and practical steps you can take to bring joy and fulfillment to your work and to the lives of others. Ryuho Okawa offers key mindsets, attitudes, and principles that will empower you to make your hopes and dreams come true, inspire you to triumph over setbacks and despair, and help you live every day positively, constructively, and meaningfully.

A Life of Triumph

Unleashing Your Light Upon the World

Softcover | 240 pages | $15.95 | ISBN: 978-1-942125-11-2

There is a power within you that can lift your heart from despair to hope, from hardship to happiness, and from defeat to triumph. In this book, Ryuho Okawa explains the key attitudes that will help you continuously tap the everlasting reserves of positivity, courage, and energy that are already a part of you so you can realize your dreams and become a wellspring of happiness. You'll also find many inspirational poems and a contemplation exercise to inspirit your inner light in times of adversity and in your day-to-day life.

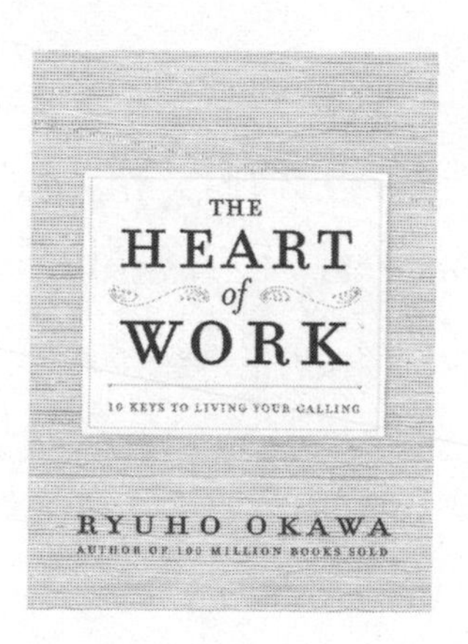

The Heart of Work

10 Keys to Living Your Calling

Softcover | 224 pages | $12.95 | ISBN: 978-1-942125-03-7

Ryuho Okawa shares 10 key philosophies and goals to live by to guide us through our work lives and triumphantly live our calling. There are key principles that will help you get to the heart of work, manage your time well, prioritize your work, live with long health and vitality, achieve growth, and more. People of all walks of life from the businessperson, executive, artist, teacher, mother, to even students, and more will find the keys to achieving happiness and success in their special calling.

Think Big!

Be Positive and Be Brave to Achieve Your Dreams

Softcover | 160 pages | $12.95 | ISBN : 978-1-942125-04-4

This self-development book offers practical steps to consciously create a life of rewarding challenge, fulfillment, and achievement. Using his own life experiences and wisdom as the roadmap, Ryuho Okawa inspires us with practical steps for building courage, choosing a constructive perspective, finding a true calling, cultivating awareness, and harnessing our personal power to realize our dreams.

THE LAWS OF MISSION

Essential Truths for Spiritual Awakening in a Secular Age

HEALING FROM WITHIN

Life-Changing Keys to Calm, Spiritual, and Healthy Living

THE UNHAPPINESS SYNDROME

28 Habits of Unhappy People (and How to Change Them)

THE MIRACLE OF MEDITATION

Opening Your Life to Peace, Joy, and the Power Within

THE ESSENCE OF BUDDHA

The Path to Enlightenment

THE LAWS OF JUSTICE

How We Can Solve World Conflicts and Bring Peace

INVITATION TO HAPPINESS

7 Inspirations from Your Inner Angel

MESSAGES FROM HEAVEN

What Jesus, Buddha, Muhammad, and Moses Would Say Today

THE LAWS OF THE SUN

One Source, One Planet, One People

SECRETS OF
THE EVERLASTING TRUTHS

A New Paradigm for Living on Earth

THE NINE DIMENSIONS

Unveiling the Laws of Eternity

THE MOMENT OF TRUTH

Become a Living Angel Today

CHANGE YOUR LIFE,
CHANGE THE WORLD

A Spiritual Guide to Living Now

For a complete list of books, visit OkawaBooks.com